From Mastery to Mystery

From Mastery to Mystery

A Phenomenological Foundation for an Environmental Ethic

BRYAN E. BANNON

OHIO UNIVERSITY PRESS ATHENS

Ohio University Press, Athens, Ohio 45701
ohioswallow.com

Printed in the United States of America
Ohio University Press books are printed on acid-free paper.∞ ™

20 19 18 17 16 15 14 5 4 3 2 1

Library of Congress Cataloging-in-Publication Data

Bannon, Bryan E., [date]–
From mastery to mystery : a phenomenological foundation for an environmental ethic / Bryan E. Bannon.
pages cm. — (Series in Continental thought ; No. 46)
Includes bibliographical references and index.
ISBN 978-0-8214-2063-8 (hc : alk. paper) — ISBN 978-0-8214-2064-5 (pb : alk. paper) — ISBN 978-0-8214-4469-6 (pdf)
1. Philosophy of nature. 2. Naturalness (Environmental sciences) 3. Ecology—Philosophy. 4. Phenomenology. 5. Environmental ethics. I. Title.
BD581.B353 2014
113—dc23

2013038523

For Mr. James D. Poisson,
whose spirit continues to inspire.

CONTENTS

ACKNOWLEDGMENTS

Expressing one's gratitude to all the individuals who have contributed in some capacity to the completion of any project of this length is a somewhat daunting task. Rather than open the usual Pandora's box of acknowledgment that accompanies the production of a book, I would like to recognize as a group all those who have discussed this project with me throughout its construction. I would, however, also like to single out a few people whose friendship and conversation about various aspects of the project have helped attune my thinking in various ways. I have in mind here David Gougelet, Lisa Guenther, Don Landes, Scott Marratto, David Morris, Alexandra Morrison, Jim Ridolfo, and John Russon.

I would also like to recognize those whose advice, intellectual generosity, and direction shaped the manuscript into its current form: Mary Beth Mader and Robert Bernasconi. Mary Beth deserves special recognition for her suggestion of what I consider to be such a lovely title for the book. Additionally, boundless gratitude is owed to Leonard Lawlor, for both his detailed analysis of a previous incarnation of the manuscript and his enduring support of me as a philosopher. Len's friendship and insights have been invaluable to my continuing to write and think philosophically.

Speaking of friendship leads me to recognize Galen Johnson and Cheryl Foster as well, both of whom taught me as an undergraduate what it is to live philosophically. My ongoing conversation with Galen, aspects of which appear in the following pages, is one of the great pleasures I have had in becoming a professional philosopher.

I also want to thank the various individuals with Ohio University Press for their assistance in bringing the book to print. Ted Toadvine, especially, for encouraging its submission and guiding me through that process; one of the anonymous reviewers whose especially insightful comments significantly improved both my line of thinking and my presentation; and Kevin Howarth, Beth Pratt, Sarah Welsch, Charles Sutherland, and (especially) Deborah

Wiseman for aiding in various parts of the production process. I very much appreciate the time and effort taken by all.

Most importantly, I would like to thank Erinn Gilson for her love, support, and tolerance of my rambling on about the history of being. This book would not be were it not for her caring and generous sharing of herself.

ABBREVIATIONS

In all applicable cases within the text, the original language citation will precede the citation of the translation. Also, due to complications involved in marshaling the technical vocabulary of philosophers in three different languages, I have frequently and without noting the fact altered the standard translations so as to clarify the interconnections between the texts and the dialogue between thinkers. I assume responsibility for any inaccuracies or uncomfortable phrasings that result from this practice.

Works by Martin Heidegger:

BP *Gesamtausgabe*. Vol. 65, *Beiträge zur Philosophie*. Frankfurt: Klostermann, 1989. Translated by Parvis Emad and Kenneth Maly as *Contributions to Philosophy (from Enowning)* (Bloomington: Indiana University Press, 1999).

EM *Gesamtausgabe*. Vol. 40, *Einführung in die Metaphysik*. Frankfurt: Klostermann, 1983. Translated by Gregory Fried and Richard Polt as *Introduction to Metaphysics* (New Haven, CT: Yale University Press, 2000).

FT "Die Frage nach der Technik." In *Gesamtausgabe*, vol. 7, *Vorträge und Aufsätze*, 7–36. Frankfurt: Klostermann, 2000. Translated by William Lovitt as "The Question Concerning Technology," in *The Question Concerning Technology and Other Essays* (San Francisco: Harper Colophon, 1977), 3–35.

SZ *Sein und Zeit*. Tübingen: Niemeyer, 1927. Translated by Joan Stambaugh as *Being and Time* (Albany: State University of New York Press, 1996).

UK "Der Ursprung des Kunstwerkes." In *Holzwege*, 7–68. Frankfurt: Klostermann, 1957. Translated by Julian Young and Kenneth Haynes as "The Origin of the Work of Art," in *Off the Beaten Track* (New York: Cambridge University Press, 2002), 1–56.

WBP "Vom Wesen und Begriff der Physis Aristoteles, *Physik* B, 1." In *Gesamtausgabe*, vol. 4, *Wegmarken*, 237–99. Frankfurt: Klostermann, 1967. Translated by Thomas Sheehan as "On the Essence and Concept of *Physis* in Aristotle's *Physics* B, I," in *Pathmarks*, ed. William McNeill (Cambridge: Cambridge University Press, 1998), 183–230.

ZW "Die Zeit des Weltbildes." In *Holzwege*, 69–104. Frankfurt: Klostermann, 1957. Translated by Julian Young and Kenneth Haynes as "The Age of the World Picture," in *Off the Beaten Track* (New York: Cambridge University Press, 2002), 57–85.

Works by Bruno Latour:

NBM *We Have Never Been Modern*. Translated by Catherine Porter. Cambridge, MA: Harvard University Press, 1993.

PH *Pandora's Hope: Essays on the Reality of Science Studies*. Cambridge, MA: Harvard University Press, 1999.

PN *Politics of Nature: How to Bring the Sciences into Democracy*. Translated by Catherine Porter. Cambridge, MA: Harvard University Press, 2004.

Works by Maurice Merleau-Ponty:

AD *Les aventures de la dialectique*. Folio ed. Paris: Gallimard, 1955. Translated by Joseph Bien as *Adventures of the Dialectic* (Evanston, IL: Northwestern University Press, 1973).

LI "Le langage indirect et les voix du silence." In *Signes*, folio ed., 63–135. Paris: Gallimard, 1960. Translated by Michael B. Smith and revised by Leonard Lawlor as "Indirect Language and the Voices of Silence," in *The Merleau-Ponty Reader*, ed. Ted Toadvine and Leonard Lawlor (Evanston, IL: Northwestern University Press, 2007), 241–82.

N *La nature: Notes, cours du Collège de France*. Paris: Seuil, 1994. Translated by Robert Vallier as *Nature: Course Notes from the Collège de France* (Evanston, IL: Northwestern University Press, 2003).

NC *Notes de cours: 1959–1961*. Edited by Stéphanie Ménasé. Paris: Gallimard, 1996.

OE *L'oeil et l'esprit*. Paris: Gallimard, 1964. Translated by Michael B. Smith and revised by Leonard Lawlor as "Eye and Mind," in *The Merleau-Ponty Reader*, ed. Ted Toadvine and Leonard Lawlor (Evanston, IL: Northwestern University Press, 2007), 351–78.

PhP *Phénoménologie de la perception*. Paris: Gallimard, 1945. Translated by Colin Smith as *Phenomenology of Perception* (New York: Routledge, 1962).

PM *La prose du monde*. Edited by Claude Lefort. Paris: Gallimard, 1969. Translated by John O'Neill as *The Prose of the World* (Evanston, IL: Northwestern University Press, 1973).

SC *La structure du comportement*. Paris: Quadridge/PUF, 2006. Translated by Alden Fisher as *The Structure of Behavior* (Pittsburgh, PA: Duquesne University Press, 1963).

VI *Le visible et l'invisible*. Paris: Gallimard, 1964. Translated by Alphonso Lingis as *The Visible and the Invisible* (Evanston, IL: Northwestern University Press, 1968).

Works by other authors:

AN Vogel, Steven. *Against Nature: The Concept of Nature in Critical Theory*. Albany: State University of New York Press, 1996.

CN Whitehead, Alfred North. *The Concept of Nature: Tarner Lectures Delivered in Trinity College, November 1919*. Cambridge: Cambridge University Press, 1920.

FI Sallis, John. *Force of Imagination: The Sense of the Elemental*. Bloomington: Indiana University Press, 2000.

FMN Plumwood, Val. *Feminism and the Mastery of Nature*. New York: Routledge, 1993.

NP Canguilhem, Georges. *The Normal and the Pathological*. Translated by Carolyn R. Fawcett and Robert S. Cohen. New York: Zone Books, 1989.

SA Leopold, Aldo. *A Sand County Almanac*. New York: Ballantine, 1970.

SCN Evernden, Neil. *The Social Creation of Nature*. Baltimore: Johns Hopkins University Press, 1992.

SS Abram, David. *The Spell of the Sensuous: Perception and Language in a More-Than-Human World*. New York: Vintage, 1996.

INTRODUCTION

THE QUESTION OF NATURE

> No important change in ethics was ever accomplished without an internal change in our intellectual emphasis, loyalties, affections, and convictions.
>
> —Aldo Leopold, *A Sand County Almanac*

> An ethics of the environment must begin with the sheer and simple fact of being struck by something wrong happening in the surrounding world. It is by noticing that something is out of joint—does not fit or function well—that a response is elicited and an action induced.
>
> —Edward S. Casey, "Taking a Glance at the Environment"

In the quotes above, Leopold and Casey agree upon a little-explored insight: environmentalism begins with a feeling. For some, this sentiment is an intuition that there is something amiss with how many human beings currently live within and interact with nature. For others, it is an emotional connection to a place or an acknowledgment that many current environmental practices are simply unsustainable. There is no one source of environmentalism that can be identified as the most important, but in all cases we *care* for nature first, and from there we consider what is to be done. For this reason, an important question that usually remains unspoken is, How do those of us already concerned with the integrity, stability, and beauty of our planet's ecological systems inspire others to care about nature? Unfortunately, such a task can appear even more daunting than convincing the government of the United States to accept a global agreement on climate change. While caring about nature is clearly not sufficient to ensure the adoption of a more ecologically friendly lifestyle, without such care I venture that environmentally friendly policy is doomed to failure. The question we face, then, is how to bring about such concern for the welfare of nature.

The problems we encounter in this regard are myriad, and there is little consensus concerning what they are and how they are to be addressed. This book will argue that the main problem we encounter in establishing a caring concern for nature is a hermeneutic one. Hermeneutics, very generally speaking, is the practice of discovering and interpreting meaning. But it does not merely deal with the meaning of linguistic expressions. Rather, modern hermeneutics finds meaning in social practices and phenomena, bodies, and even existence itself. Central to hermeneutical practice is the notion of an interpretive framework: human beings seek and discover meaning only within the context of a specific network of already established meaningful relations. To assert the existence of an interpretive frame is not to say that the frame cannot be shifted or altered. In fact, the claim that I will make is that a change of precisely that kind is what is necessary: we must alter the very way nature is conceived in order to inspire the necessary changes in affection that can motivate a healthier relationship with nature. Speaking negatively, without such a shift we lack a solid normative framework from which to voice our environmental concerns.

Bryan G. Norton has already indicated this problem as the "environmentalists' dilemma" in his *Toward Unity Among Environmentalists*.[1] On the one hand, the language and meaning that dominates policy discussion is that of economics, and that language does not capture the true force of environmentalists' concerns. Saying, for example, that mountaintop removal is reprehensible because it fails to include in its economic evaluations supposed "externalities," such as the poisoning of rivers and the destruction of local communities, fails to capture the indignation and dismay that many feel as a result of the practice. On the other hand, environmentalists have failed to develop a common tongue to express their concerns, so the economic lingua franca remains unchallenged. Since many proponents of preservationism (though by no means all) believe that the economization of value is one of the causes of the domination of nature while the conservationist is less averse to such thinking, the two groups will from time to time end up fighting each other rather than opposing practices that threaten their commonly held view of a more ecologically responsible lifestyle. Norton proposes resolving this dilemma by reconciling the conservationist and preservationist approaches to environmentalism, not by means of dealing with what he calls their "two apparently exclusive worldviews and sets of value assumptions," but by focusing on their shared objectives.[2] Since his goal is policy-oriented, this decision makes sense to a certain extent, but for those of us who are made uncomfortable by the implications of the worldview used to support both the economic/conservationist and the preservationist positions, Norton's solution fails to touch the problem he so astutely identified in the first place: the very ways in which different people conceive of nature lead us to identify significantly different practices as environmentally friendly or destructive.

So while Norton may be correct in indicating the common ground held between members of different camps of environmentalists, the sad fact is that in many cases consensus on how best to deal with these problems deteriorates as soon as a plan of action begins to be discussed. Take, for an example, the practice of ecological restoration. Though practitioners tend to have good intentions, some environmentalists have levied criticisms against the practice: that it indicates a resignation of the main goals of preservationists; that it is impossible to restore nature since any system that we could reestablish would be an artifact; that restored systems require continual management to remain in their restored state; that the project itself reflects a romantic nostalgia for things past, and so on. Against these charges, those who defend the practice maintain that it is one of the best ways to expand the wild spaces we have remaining and to repair damages wrought in the industrial era. I do not intend to mount a defense of ecological restoration here, but I do wish to note the manner in which the interpretive frameworks of the two sides contribute to an impasse of sorts. Opponents tend to view nature as an order devoid of human presence, while advocates see nature more in terms of a cybernetic system. Even in this latter group, however, there are those who view human influence, or at least that of Europeans and their descendants, as inimical to nature, which is why the pre-Columbian period is usually selected as a target for restoration. There are environmentally conscious individuals in both camps, but the differences in their interpretive framework prevents them from perceiving the situation in the same light. This divergence of perspective is a result of the interpretive framework each employs, which is constituted by a certain set of assumptions *of a metaphysical character*; what is in question is *what nature is* and therefore *what is natural*. The dispute seems to revolve around the legitimacy of a practice, but the conflict is actually situated at the level of meaning. Hence the need for hermeneutical intervention.

John van Buren has seen the same need and has therefore proposed pursuing a critical environmental hermeneutic. He describes such a hermeneutic as occupied primarily with "the *sense* or *meaning* of the environment for perceivers and is thus unlike the natural sciences, which are focused primarily on the biophysical aspects of the environment."[3] Such a hermeneutic would develop along three distinct axes: "(1) environmental epistemology (describing and critically evaluating the different views of what the environment is), (2) environmental ethics in a narrow sense (describing and evaluating views of the value of environment), and (3) environmental politics (describing and evaluating who has or should have political power in the environment)."[4] Importantly, however, van Buren notes how these questions involve one another: narratives concerning what the environment is will be laden with value, both ethical and political.[5] Though his essay deals primarily with how these questions pertain to forests, they also suggest a broader project, namely how best to interpret and understand the narratives that structure our metaphysical understanding of nature itself. My intention in writing this book is to undertake exactly this project: to provide a hermeneutic of nature that will enable us to approach the ethical and political problems that we face concerning the natural world in a different way. Unlike van Buren, then, my aim in discussing the narrative structure of the conception of nature is not merely descriptive, though I agree with him strongly concerning its potential to resolve vexing political disagreements concerning the environment.[6] Rather, with the insight that these conceptions are in fact both ethically and politically normative, I am arguing for a way of thinking about nature that can help us to address the pressing concern of how to make sense of individuals' perceived indifference to the environment (at least in terms of their behavior) and perhaps impart some sense of why it is important to care for nature. My subsidiary concern is to avoid placing us within the event horizon of certain philosophical black holes (e.g., the debates concerning intrinsic value) in environmental ethics.

Such a project is not without precedent.[7] Val Plumwood, perhaps more than any other philosopher, has shown us how the hermeneutical concerns about the significance of nature and our ethical intuitions are intertwined. In *Feminism and the Mastery of Nature*, she analyzes the ways in which our metaphysical beliefs concerning nature have been used to justify and promulgate politically oppressive forms of behavior.[8] In her view, the common root of the oppression of women and the domination of nature lies in dualistic thinking. Dualisms operate along five key dimensions: (1) the creation of two *wholly distinct* classes on the basis of a difference in properties (hyperseparation); (2) defining the identity of the disempowered class in terms of properties it lacks in relation to the empowered class (relational definition); (3) the instrumental objectification of the disempowered class; (4) a denial of the dependency the empowered class has upon the disempowered (backgrounding); and (5) stereotyping members of the disempowered class so as not to recognize salient differences between its members (homogenization).[9] As should be clear, dualisms differ from dichotomies in that they establish a hierarchy that licenses the oppression or objectification of one of the classes. In order to secure the political order they underwrite, dualisms also typically form networks (e.g., the relation between reason-emotion, culture-nature, and man-woman) (*FMN*, 44–47). The nature-culture binary is a particularly pernicious dualism at the foundation of much Western thinking. Plumwood traces the form of this dualism beginning with Plato's thought into our contemporary metaphysics, revealing the different ways in which the philosophical tradition is party to establishing a metaphysical discontinuity between humanity's cultural and historical life and "the rest" of nature. Descartes's metaphysical dualism provides a convenient example: Bodily substances are hyperseparated from mental substance as a result of defining natural bodies in terms of properties they seem to lack in relation to minds. The body is then reduced to the instrument of the mind's agency, and its contribution to that agency is discounted as being inessential to knowing. Animals, even as they display certain intentional properties that are allotted to the mind alone, have those differences with objects such as rocks effaced in an effort to homogenize them within the category of natural bodies. The legacy of Descartes's dualistic philosophy continues to shape the contemporary philosophical landscape insofar as current forms of "naturalism" accept the Cartesian understanding of matter and strive to explain phenomena such as agency and intentionality in terms of supposed material processes.[10] If the domination of nature takes this dualizing form, then the first step toward liberating nature is the hermeneutical project of developing a new sense of nature. To do otherwise is to persist in relating to nature as a passive domain radically different from ourselves and preserve the dualism of nature-culture.

Though I agree with Plumwood's analysis, the issue is made more complex by the fact that other philosophers who share the same diagnosis offer an alternative solution to the problem: rejecting the concept of nature in its entirety rather than reformulating it at all. Neil Evernden, for example, also argues that the assumed dualism between the two is a philosophical artifact that serves to underwrite a political scheme.[11] In his view, the concept of nature is just as easily employed to justify progressive agendas as it is oppressive ones, which is why naturalism cannot serve to justify or explain plans for social action (*SCN*, 16–17).[12] To demonstrate his point, he presents the debate we have been discussing between those who "wish to husband nature efficiently" and those who "wish to treat it well, as one would (or should) another person" as one that rests on rather shaky foundations (101). Both sides accept that there is something that stands in the place of nature, but they define the concept in such a way that it supports their political and ethical goals. For this reason, Evernden takes nature to be a placeholder concept that embodies the agenda of the group in question. So conservationists tend to define nature in a materialistic fashion, while preservationists tend to want to find something *more* in nature (e.g., intrinsic value, spiritual meaning, or some larger self). The dispute is irresolvable because *nature is not a thing with properties*. Thus, neither the conservationist nor the preservationist has a valid position because they both make the mistake of thinking of nature as a thing. In Evernden's view, the best course of action is to cease speaking of nature and to approach each being in its novelty.

So both Evernden and Plumwood agree that the concept of nature is inherently linked to the establishment of a particular social order. In accepting that connection between the two, the divide between ontology and ethics erodes: responses to metaphysical questions contain within themselves an inherently socially oriented practice, even in science. This relationship between nature and politics is what prompted Arne Naess to advise that philosophy supporting environmentalism "move from ethics to ontology and back."[13] In following this advice, we begin with ethics because it is built upon the shared sentiment that we ought to care for nature, proceed to ontology to help us determine how and why we ought to care, return to ethics in order to seek a way to enact that program, and always remain open to further ontological investigation that can shift us from our current path. For the purposes of the current work, I assume an ongoing ethical commitment to a healthier relationship to the environment and begin with the second step of the process, namely an ontological investigation into the character of nature. I believe this is fair given that all the parties I will discuss orient themselves toward establishing a nondominating relationship to other beings. My goal is to preserve the intuitions that sustain both Plumwood's and Evernden's analyses, specifically that the way forward for environmentalism lies along the road of a reformulated conception of nature, but at the same time that it would be a mistake to think about nature as a substantial being that bears properties.

Retaining the concept of nature is critical for at least two major reasons. First, there is already an emotional connection between many people and their environs that is clearly expressed in the term. Even if the intention is to change the relationship between individual human beings and the rest of nature, it will be easier to enact that change by asking people to focus on an already understandable and acknowledged relationship rather than by asking them to reconceive their experience as a whole from the very beginning. Though the latter reconception may be the ultimate goal, we must, as Martin Heidegger would say, "twist free" of our already established ways of thinking before we can think anew. The second reason to continue to speak of nature follows from this point, namely, that the only way to understand and repair the damage we have wrought within the natural world (and one another by means of various strategies of naturalizing oppressions) is by cultivating a more progressive conception of nature (*FMN*, 195–96). As Plumwood makes this case in "Toward a Progressive Naturalism," "the fact that a few people have begun to contest the devaluing and agentic disappearance of nature or woman does not mean that we have arrived at a system of thought or life that can dispense with the concept."[14] In other words, because the dualistic conception of nature and humanity has been the means by which various oppressive political schemes have been justified and perpetuated, asserting counternarratives that challenge the dualizing master narrative constitutes the first step in twisting free of the oppressive aspects of the metaphysical inheritance of the West.[15] These narratives ought to focus on establishing the ontological continuity between humanity and nature, while also preserving the specificity and difference of each constituent of nature.[16] The importance of this strategy lies in how it simultaneously challenges dualistic thinking via the ontological continuity while not homogenizing beings and thereby allowing what is distinct about each to shine through.

A more vexing problem for the current hermeneutical project emerges, however, when we return to the ethical domain and ask what is to be done regarding the human impact upon nature. One of the many consequences of offering a counternarrative to the dominant conception of nature in Western thought is that what it means to preserve such a nature is going to change radically. Preservation usually refers either to sustaining the ability of natural systems to self-regulate or to keeping what is left of so-called "pristine" nature undespoiled and beyond the reach of humanity. On this view of preservation, the normative status of nature is taken as given: human actions ought to be governed by what is "good for nature." Calling for preservation requires, in at least a limited sense, taking the current state of nature to be more valuable than other states; it requires a normative evaluation of the current state of affairs with respect to a possible future. Given the current state of environmental affairs along with the political paralysis that impedes changes in that state, such calls for preservation are understandable, but on what basis is the value judgment made? There is a need, then, to determine the basis of the normative claim. Everything hinges upon *why* nature should direct human action and *how* one comes to determine what is good for nature.

Philosophers have, of course, provided responses to these two questions, but, as Steven Vogel points out, these responses frequently lead to antinomies that make the responses untenable.[17] The problem stems mainly from the competing commitments that philosophers sympathetic to the preservation of wilderness want to maintain. On the one hand, there is a strong commitment to antihumanism or, as it is stated more usually, a resistance to anthropocentrism. For example, Tom Regan argues that any ethic that would call itself an environmental ethic, in order to distinguish itself from traditional ethical systems that deal with relations between human beings, must accept that there are both nonhuman beings and some nonconscious beings that have "moral standing."[18] The justification for his claim is that if one does not accept either of these conditions, the formulated ethical system will be one for the use of the environment rather than an environmental ethic.[19] Additionally, if one accepts that nonhuman animals are morally considerable but not at least some nonconscious beings, then one is advancing a "kinship ethic" based on the extension of human ethics to certain other animals that exhibit similar properties rather than formulating a distinctively environmental ethic. Regardless of whether one accepts the moral framework within which Regan operates,[20] his point that what makes an environmental ethic unique is an antihumanistic standpoint is an important one because it underlines the central problematic of the line of questioning that calls itself environmental ethics: What place does humanity have within the natural world? Is there a responsibility, or perhaps even a duty, to *manage* nature? Is nature still nature if it is placed completely under the management of human beings?

The disquietude that some environmental thinkers feel with respect to the management of wild places brings to the fore another commitment involved in arguing for the preservation of wilderness, a belief in the inherent "freedom" or "independence" of nature. Usually, this freedom simply refers to the fact that nature is capable of functioning on its own without human intervention. The hope, however, in making such a claim seems to be that in affirming the independence of nature from humanity, nature will be acknowledged as having a domain beyond or outside that of human action worthy of preservation. By setting up nature as a sphere apart, one can proceed to argue that it ought to remain outside the influence of human beings, remain in an untouched state. Even Plumwood, who criticizes this idea as another embodiment of a hyperseparation of humanity and nature, still argues for the autonomy of nature as embodied in a particular kind of teleology (*FMN*, 134–36, 161–64).

Aldo Leopold's "land ethic" serves as a good example of how the commitments to antihumanism and to nature as an independent sphere work together. In many ways, Leopold's "land ethic" is still a dominant voice in environmental ethics, since his work serves as a touchstone for a wide range of disciplines. The centerpiece of the theory is the antihumanistic assertion that humanity is a "plain member and citizen" of "the land-community" (*SA*, 240).[21] The land, for Leopold, is a way of expressing the ecological system of energy exchange that occurs between bodies of different types. So the observation is that humans, just as all other natural beings, are subjected to and dependent upon ecological laws. As is obvious throughout the text, however, Leopold conceives of the land in terms of an organism, in other words, as an independent, self-regulating being (e.g., 190, 195, 261, 272). What is relevant for our purposes is that the land ethic finds in the processes of nature a form of organized independence that leads him to break with ordinary conceptions of the land as property and conceive of it as a "biotic mechanism" (251). With the antihumanistic insistence on humanity's being just a plain member of the land community, combined with the insight that nature forms an independent realm of circulating energy, Leopold asserts his one normative principle that transforms his ethic into a moral theory: "A thing is right when it tends to preserve the integrity, stability, and beauty of the biotic community. It is wrong when it tends otherwise" (262).

Together, then, these two commitments provide the foundation of arguments for the preservation of nature and humanity's ecological place *within* the larger system of nature rather than with respect to it. However, in the convergence of these two commitments, Vogel identifies a tension that pervades much of the philosophy that deals with nature. If one accepts antihumanism, then one accepts that humanity is just another part of nature, as are other animals, plants, and stones. This point is readily seen in Leopold. To do so, however, requires that the effects of human behavior be considered just as natural as the effects of the behavior of any other living being, since the difference between the two sets of actions is one of magnitude, a point with which Leopold also agrees (*SA*, 255–58). Given this situation, if the normative foundation for an environmental ethic is to be what is good for nature, then it is difficult to see how one is to criticize natural behavior as contravening nature. Leopold's principle seems to break down at the epistemological level: How does one determine what does and does not preserve nature without, in a certain sense, placing oneself outside of it? After all, on this view a humanized nature is still nature. In a very general way, we might say that certain practices, such as dumping toxic chemicals into a river, are inherently bad, but only by affirming a certain stability and fixity of nature that would make it susceptible to "wise human management" and thereby undermine the value of the independent evolutionary processes of nature. If one desires to criticize human behaviors as violating or going against the integrity of nature, placing the emphasis on upholding the independence of nature from humanity, then, in this context at least, doing so requires the reassertion of human separatism due to the fact that innumerable animal (e.g., beavers or elephants) and plant (e.g., kudzu or purple loosestrife) species act against the integrity of their environments and yet we do not see anything "wrong" with their actions. In some cases, such actions are even used as evidence against their rationality. To criticize humanity for similar actions results either in reneging upon the commitment to antihumanism (we should "know better," which implicitly affirms the humanist belief in exceptionalism) or in slipping into the misanthropic attitude of which Murray Bookchin has accused deep ecology (we are like kudzu, choking out life wherever we go).[22] Without further elaboration, then, the two main commitments of environmental ethics oriented toward the preservation of nature seem to be incompatible.

Vogel's solution to these difficulties is also to abandon the concept of nature in favor of an "environmental philosophy without nature."[23] What he intends to do is to overcome the tensions in arguing for preservation through the acknowledgment that human beings transform their environments, just like any other living being, and so nature is a social construction.[24] When Vogel speaks of social construction, however, he has a very precise sense for the term in mind. Rather than the type of social construction of *ideas* that is usually discussed, Vogel argues that one can explain the current state of the global environment by accepting that it is a product of human actions, a work of human labor, even if unintentionally. To make such an assertion, however, is to deny directly the second commitment discussed above while at the same time acknowledging at least a significant portion of humanity's responsibility for the current state of affairs (*AN*, 10). While the result of denying the independence of nature allows us to hold humanity accountable for the environment, it also places nature entirely under the aegis of the human, social realm (166). Subsuming nature into the social realm removes the problem of establishing nature as the source of normativity because nature, as a product of human action, has a value derived from its value for human beings. It also eliminates the problem that concerned Evernden of situating the source of normativity in a concept that lies beyond the reach of rational contestation (9–10). Environmental ethics, then, is no different from other forms of political and ethical questions; it simply deals with a particular genre of human interaction: humanity's communication with its environs. "Environmental ethics" should thus be heard quite literally: what is under discussion is the status of the human environment, not the health of nature. I use the words "communication" and "discussion" quite deliberately given that Vogel is expanding Habermasian discourse ethics so as to make it better suited to deal with questions involving at least one party that does not engage in discourse. While nature cannot, in Vogel's view, enter into dialogue, and thus also cannot be a source of values or have considerability independent of its relation to human speakers, one can make ethical claims concerning the environment precisely because it is an object of human discourse and culture (170). Of course, this brings about a rejection of the commitment to antihumanism as well. But since Vogel does not limit himself to instrumental rationality, relying instead on Habermas's notion of communicative rationality, the value found in nature is not purely one of what is *useful* to humans. Rather, he states, "the question of value in nature, such a theory suggests, is a question to be answered by a *nature-building process become self-conscious*, one that knows the world it inhabits to be its own construction and its own responsibility and so poses for itself—discursively—the question of what that world should be like" (171). To return to the earlier example, the reason we do not dump toxic chemicals into the river is just that most responsible citizens would agree that a world with toxic waters is not a desirable one to live in. The vision, then, that is to replace the preservation of nature is one of humanity fashioning the surrounding physical world into a suitable environment for itself and other living beings.[25]

While such a view certainly would serve to address the charges of misanthropy levied against those who would preserve nature at the expense of humanity, it does so at the cost of what many believed was meaningful about pursuing an environmental ethic: its antihumanism. Vogel's concerns about social justice and the coherency of an environmental ethic as currently framed, however, are well grounded, and so we are faced with another challenge: Is a distinctively ecological ethic possible? Can one maintain the two competing commitments that seem to be required for such an ethic without leading to incoherence? The second major concern of this book will be to formulate a response to this challenge in the affirmative. The strategy for doing so will be to argue that Vogel dismissed the thought of nature too soon: in developing a nondualistic counternarrative regarding nature, resources emerge for making normative claims about nature without falling prey to the pitfalls Vogel astutely identifies. If one begins from the intuition that the anthropocentric worldview is at least partially responsible for the current state of affairs in the environment, an intuition that many environmental philosophers share, Vogel's attempt to avoid the tensions inherent in environmental preservationism is somewhat of a nonstarter.[26] Luckily, he leaves open a route for our hermeneutic project to provide an alternative resolution: if, as he affirms, an environmental ethic's shortcomings are derived from a need for "a *considerably* more sophisticated account of what it means when it speaks of 'nature' than has in general heretofore been forthcoming if it is to achieve any important results" (*AN*, 8), then there is at least some hope of formulating an authentically environmental ethic in the traditional sense once we arrive at a reformulated conception of nature. For current purposes, then, Vogel should be understood as providing a provocation to a philosophy of nature, not a decisive blow.

To recapitulate briefly what has been argued so far, we can identify a need for a hermeneutic of nature coming from two distinct sets of concerns. The first set is the practical problem of how to create an affective relationship to nature in order to engender a nondominating relationship with the nonhuman world. The second emerged from the projected response to this need and concerns the danger of accepting conflicting guiding principles once we begin to construct an ethic based upon a certain conception of nature. In both cases, I have suggested that by examining the dominant philosophical conceptions of nature within the Western tradition and offering an alternative way of understanding what nature is and how it operates, we can begin to address these legitimate and pressing concerns. For this reason, what I offer here should be viewed as a *foundation* for a normative theory, not in and of itself a work that is going to offer any concrete answers to environmental problems (however they are formulated).

I have chosen the phenomenological tradition as a means to elucidate the alternative conception of nature for a number of reasons,[27] the first and foremost of which is the extent to which those thinkers have rigorously maintained that our understanding of nature is a faulty one and in need of correction. In many ways, the problem of naturalism *is* the problem of phenomenology.[28] Thus, attempting to give some account of a phenomenological conception of nature is a convenient way of developing a novel conception of nature from within the philosophical tradition rather than from outside of or opposed to it.

Each of the chapters reflects the above intention in a different way. I begin in the first chapter by dealing with what I regard as the most serious of contemporary attempts to eliminate the discussion of nature from environmentalism. That challenge emerges from the work of Bruno Latour and some of his advocates' claims that a suitable relationship to the environment can be established only once the concept of nature is relinquished. Latour's views have been mostly popular in the social sciences, but that is to philosophy's detriment. Like Plumwood and Evernden, Latour rejects the dualizing tendency in modern thought that gives rise both to the dominant forms of contemporary naturalistic philosophies and to the schools of social construction that seek to oppose them. The problem with nature is that we usually understand it as simultaneously possessing immanent laws that establish the relations between beings in advance and from the start and transcendent of humanity such that it should remain independent of any "pollution" from the social world. Even if a philosophy is not outrightly dualistic, these dualistic presumptions continue to color our relation to nature, and so the concept ought to be discarded as a relic of the unattainable modernist ideal of total purification of nature from the social and of the social from nature. He further blames the modernist conception of nature for giving rise to a monolithic "Science" (rather than "the sciences") in order to know it, which creates profoundly undemocratic institutions and practices. Leaving nature behind allows us to have a more charitable (not to mention accurate) understanding of nonhumans, agency, and the work of the sciences, which in turn he believes will give rise to more democratic institutions. Such institutions are necessary to having a healthier relationship to the other members of the common world that we all inhabit together. In establishing these points, Latour's analysis is rich and insightful, which is precisely why it needs to be thought through.

Since Latour considers himself an opponent of phenomenology, and Martin Heidegger's version in specific, the second chapter begins with a consideration of how their two views can be seen as complementary and the limitations of their alliance. In order to elucidate these matters, I begin by explaining Heidegger's "history of being," in which some others have found resources for contemporary ecologically minded philosophy.[29] Heidegger offers an explicit critique of the popular ecological conception of nature as an organized system, arguing that such views are a product of a philosophical evolution that begins with pre-Socratic thinking. Though I make no attempt to reconstruct Heidegger's argument concerning the history of being as a whole, significant time is spent developing the implications of conceiving of nature as the central conception of all Western philosophy. Heidegger's intent here is akin to Plumwood's and Evernden's insofar as all three are striving to demonstrate how certain conceptions of *what* beings are come to institute fixed, hierarchical ways of dealing with those beings. Like Foucault after him, Heidegger conceives of this kind of systematic ordering as a form of domination.[30] For Heidegger, what makes this domination possible is mistaking a given interpretation of the being of beings for the unchanging essence of being as such. In other words, we take a given hermeneutical understanding of the world and objectify it such that it becomes the *only comprehensible* hermeneutical frame for common sense. Mastery, then, is a hermeneutical issue: beings are dominated by reducing them to any one of their appearances. In the case of nature, the forgotten assumption is that a being's essence is defined in terms of what is constantly present, what is invariant. What rectifies this situation is a renewed attentiveness to the ways in which human beings both dwell within an inherited network of received meanings and relations (the "world") and simultaneously are always potentially authors of a future divergent from the one opened up by that inheritance. What Heidegger uncovers through his meditations is that the domination of the earth is a product of a failure to remain mindful of the essential openness of humanity and the manner in which beings are always already understood in terms of a world. The domination of beings that leads to their being reduced to their utility within a given nexus of worldly relations can be ended only by means of a continual holding open of the moment of interpretation so as not to reify any one meaning as the unchanging essence of things.

The implications of this for ecological philosophy are myriad because, in Heidegger's reasoning, the devastation of the earth that it seeks to prevent is a direct product of the domination of beings. In the third chapter, then, we will investigate how Heidegger attempts to reconceive nature as that which resists disclosure within a given hermeneutical framework. This notion, while it completely eliminates nature as a domain on its own, is the first step toward retaining the commitment to nature's autonomy because it explicitly affirms the impossibility of total human control, of subsuming nature completely within the human world. To this end, Heidegger also shows how nature and history are not incompatible and in fact must be thought together. What Heidegger offers, then, is the possibility of a relational rather than a substantial conception of nature that eschews dualistic thinking.

Heidegger, however, only shows the way to this renewed nature that is no longer dominated by humanity. From the perspective of the current project, the shortcomings of his views on nature lie in his twofold humanistic insistence: first, that only humanity has a world and therefore that language discloses only to humanity; and, second, that the processes of temporalization (and therefore of history and nature) he discusses emerge only within human existence. In moving to the fourth chapter, then, we will seek a way to overcome these difficulties while maintaining the essential insights concerning what we must seek in a conception of nature.

My proposal is that such a conception of nature can be found in the structures presented in the phenomenological ontology of Maurice Merleau-Ponty. What Merleau-Ponty offers that is not found in Heidegger is an explicit means of decentering humanity from its privileged position as the sole projector of meaning. But it was not always the case that Merleau-Ponty was driven by Heideggerian concerns. Consequently, we will trace the contours of the relational concept of nature by pursuing the development of his thought from a modified Husserlian phenomenology to a novel approach to a Heideggerian mindfulness of being. What we will find is that Merleau-Ponty's explanation of "flesh" as a foundational relationship between beings permits us to speak of nature as something other than a social construct while at the same time discussing how humanity participates in the larger whole. Though Merleau-Ponty is quite aware of scientific accounts of nature, he discovers an alternative to reductive materialistic accounts of nature in the process views of Alfred North Whitehead. The process view of nature allows him to maintain his preferred emphasis upon perception while at the same time finding a basis for a relational ontology that affords a foundational role to all bodies and other forms of life. In this view, nature is no longer thought of as a substantial domain, but as a collectively expressed sense in which all beings participate in the expression. The advantage of thinking of nature in this way is that it allows for the kind of political discussion that Plumwood, Evernden, and Vogel want to open up, while at the same time providing a vision of nature that can support a distinctively ecological ethic. So, by the conclusion of the book, we will have arrived at a conception of nature as an evolving pattern or mosaic, but one that is mutualistic—it is a common production of all beings—without being teleological. In many ways, this view of nature shares many features of Latour's "common world," a term Merleau-Ponty uses himself.

Before beginning this task, I would like to say a few final words with respect to possible ways in which the current project might be misconceived. The first possible misinterpretation is that it is taking a side in the extant debate between so-called realists and social constructivists. Like Ted Toadvine, I believe the power of the phenomenological position lies in rejecting the very terms of the debate, not in affirming either side.[31] So while the reformulated conception of nature will reflect a criticism of what I take to be a rather narrow conception of "the real" on the behalf of realists, I also will avoid making meaning and value the sole purview of human beings. The view of nature I develop here can affirm the best intuitions of both sides without having to assume either position. On its basis, we can acknowledge the independence of nature without resorting to scientific realism to do so, but we can also find ways to refer to nature as movable normative ground. So, in turning our attention to the *concept* of nature and the narratives we use to give meaning to the world around us, let us keep in mind that the fact that nature is a historical concept does not necessarily undermine the independence of the natural world. One can accept that "nature" is a human signification, subject to historical changes, while at the same time agreeing that there is a nature in which human beings are but one participant.

The second potential misconception is one based in the style of what follows. Especially in the second and third chapters, I will be dealing quite closely with texts from Heidegger and Merleau-Ponty. Though I hope that scholars of those two figures will find things of value within those pages, it is not my intention to offer a comprehensive interpretation of their respective philosophies. Rather, my intention is to draw attention to certain aspects of their thought in light of how they might be productively employed so as to address the problems described above. At times this will result in the omission of critical concepts for the philosophy overall in favor of highlighting less obviously important ones (I do not, for example, emphasize the role of poetry in combating dominating ways of thinking in Heidegger); at other times familiar concepts will be pushed in new directions in order to be more helpful in an environmental context (as with Merleau-Ponty's notion of flesh). I should also note that I have retained the technical language of these philosophers in discussing their ideas, though I have done my best to introduce and explain the concepts so as to make them more familiar to those who have not devoted themselves to understanding phenomenology.

With those caveats, the time has come to begin our inquiry. For at least four hundred years, the ethical ideal of mastery has guided the human relationship to nature. The project of mastery over nature has been embodied in diverse forms, ranging from technological exploitation, to the pursuit of absolute knowledge, to a tendency to shape and mold one's body obsessively, and so on. While I have no illusions concerning my ability to displace that ideal definitively in what follows, I do hope to shed light on another way of thinking about how to oppose it. By focusing upon the interplay between our ontological concepts and the resulting hermeneutical meanings we find in the world, the mystery of our relationships with other beings once again can become the subject of wonder and enchantment within the world. If we are to resituate the instrumental attitude toward beings that others have so rightly criticized into a more appropriate domain, a good place to start is in the rediscovery of what in the world is mysterious and necessarily remains so. My argument will be that an ontology that makes mystery an ineliminable aspect of our relationship with the world supports dispositions and attitudes toward nature, such as openness, adaptability, and caring, that are either simply not possible or unlikely under an ideal of mastery.

CHAPTER 1

THE PROMISE OF A COMMON WORLD

> We have exchanged masters many times; we have shifted from the God of Creation to Godless Nature, from there to *Homo faber*, then to structures that make us act, fields of discourse that make us speak, anonymous fields of force in which everything is dissolved—but we have not yet tried *to have no master at all*.
>
> —Bruno Latour, *Pandora's Hope*

"To have no master at all": such is the promise of the common world in the work of Bruno Latour. Living as we do under all sorts of transcendent authorities, even if it remains unrecognized that such is the case, this promise is a long way off. One may even wonder if an existence without such authority is possible. After all, without some higher values to guide decision making, it seems as if we flounder in arbitrariness. The challenge is to seek a foundation from which normative position-taking is possible without also asserting the unquestionable authority of that foundation. As we also see in the above quotation, Western thought has characterized that foundation in myriad ways, with nature remaining one of the most potent of these authorities.

Latour's thought offers one of the most systematic inquiries into the limitations of the concept of nature and what might fill the void left behind once nature is removed from its place of privilege. For Latour, nature is one of the two organizing conceptual categories of modernity. Along with the social or society, the category of nature serves one of the two ontological categories of the modern world. In modernity, everything that occurs or exists is thought to originate from either nature or society. The modern world, then, is inherently dualistic, a situation that underwrites the worst kind of political oppression, of both human and nonhuman beings. In order to create a truly democratic world, we need to separate ourselves from this dualistic way of organizing beings and understand them in terms of the kinds of relations into which they are capable of entering. By doing so, nature will no longer be conceived of as the fixed and stable background laws that determine how beings enter into causal relations governed by those laws, but rather will be just the set of objects studied and accepted by the sciences into what Latour calls "the collective."

Within this brief gloss on his system, there are many possible points of misunderstanding. In order to progress in developing a critical hermeneutic of nature, we must be careful to avoid such confusions, for to do so would be to become entangled in the morass of mischaracterizations of his work (e.g., that he believes there is no external reality). Because Latour ultimately removes nature from its position as an ontological category in order to resituate it as a political designation, his philosophy provides the best road map regarding the dangers any ontological conception of nature faces. Hence, in being attentive to the contours of his thinking, we will develop a clearer sense of what kinds of conceptualizations will merely lead us back into dualistic thinking and what kinds may lead us out. Additionally, Latour's philosophy analyzes quite clearly the interrelationship between the sciences, politics, and ethics in ways that will be illuminating if we are to use the environmental hermeneutic in a normative and not merely a descriptive fashion. Like Steven Vogel, Latour believes that "nature is the chief obstacle that has always hampered the development of public discourse."[1] So, in following Latour in his attempt to open up public discourse while retaining a place for nature in that conversation, we need to clarify why he believes the attempt to stop the domination of nature is a "reactionary" project and begin to articulate why retaining the concept of nature is neither "antimodern" nor "modern" in his sense of the terms.[2]

The majority of the argument that substantiates this last claim will be developed over the remainder of the book. Consequently, this chapter has two more modest ends: (1) to understand the potential pitfalls of retaining the concept of nature; and (2) to justify the need for a critical environmental hermeneutic in light of Latour's views. Again, the form the concept of nature will take will be most fully developed only in the fourth chapter, which deals with Merleau-Ponty's ontology. I will begin this task by explaining Latour's critical framework so as both to show the validity and promise of his concerns and to clarify the critique of metaphysical dualism for those who are unfamiliar. From there, we can delve more specifically into how the concept of nature operates within this framework, how it relates to the artificially hyperseparated political domain, and why he believes it necessarily sanctions an oppressive set of political relations. At this point, when the notion of nature seems to be discredited, Latour reverses course in order to accord a place to nature within the nonmodern world. Hence, the second section of the chapter turns toward the role that Latour retains for nature. In following this course, I will argue that Latour's philosophy is simultaneously the largest challenge that a phenomenological conception of nature will face as well as its surest guide. In order to conclude the chapter, then, I will briefly examine and preliminarily respond to his critique of phenomenology as a launching point for the remainder of the project. Though I am unsure as to whether he would be comfortable with such a characterization, I ultimately conclude that Latour, a staunch opponent of the metaphysics of nature that he associates with projects of mastery, can be enlisted as one of the strongest allies in formulating a conception of nature that can ground an environmental ethic without resorting to the metaphysical naturalism upon which so many environmental ethicists rely.

I. NATURE AND THE MODERN WORLD

In order to understand Latour's reasons for rejecting the concept of nature, it is helpful to begin by contextualizing his comments in terms of his larger philosophical project and revaluation of the worth of modernity. Alluding to his earlier *We Have Never Been Modern*, Latour summarizes his approach not as a radical project of dissolving the traditional categories of philosophical analysis within which we operate, but rather as an appeal to those who would engage in political ecology. He consistently recommends that the best way to undermine the project of modernity is to *slow down* and reveal its workings rather than undertake any more-radical project of rejecting its concepts and categories wholesale.[3] Slowing down, unearthing the processes and beings that constitute the world around us in the way we experience it, will be more effective in creating a more democratic world since one can then focus on "undoing some of the problematic relations without having to recast the whole set" (*PN*, 3). Even as the project of modernity is continually rejected throughout Latour's work, his relationship to it is rather nuanced since he ultimately does not reject its categories.

Making the extant critiques of dualistic thinking more complex, Latour characterizes modernity as the result of a "double separation," or what we might call a "doubled dualism." The first dualism is the one already discussed in terms of a hyperseparation between nature and society/history/culture (*NBM*, 13). These two categories function as ontological domains into which every being can be filtered. Even those phenomena that seem to be composed of elements that are neither wholly natural nor wholly cultural can be subjected to an operation of "purification," through which they are analyzed into natural or cultural "parts" and thereby understood in terms of the dualistic categories. The second dualism is between the two practices made possible by the hyperseparation of the human and nonhuman domains: the aforementioned process of purification and the act of translation (13). Translation, the apparent opposite of purification, is the practice of constructing hybrid networks of relations, that is, bringing together supposed natural and social elements in a single phenomenon or practice. Consider, for example, something like climate change. On the one hand, climate change can be analyzed into those natural processes that would be at play in the absence of human influence (e.g., carbon cycling or the albedo effect) and those cultural practices that contribute to altering those processes (e.g., a consumeristic economy, which is itself a very diverse phenomenon). We thereby try in the process of purification to isolate the social *causes* of climate change and reverse their influence on the natural processes. On the other hand, climate change is also irreducible to either of these sets of processes and therefore can be effectively understood only in terms of the interrelationship between social practice and natural process. In other words, social practices do not cause the natural processes to be perverted, but the agglomeration of the diverse constituents of the larger phenomenon all contribute to the way the network functions; what we call climate change just is the agency of the network. So the modern metaphysical model simultaneously encourages at least two different and opposed frameworks for analysis, the dualistic metaphysical categories and the contradictory hybridizing of those categories.

This doubly dualistic situation is instantiated in what Latour calls the "paradox of modernity." In short, the paradox is that "if we consider hybrids, we are dealing only with mixtures of nature and culture; if we consider the work of purification, we confront a total separation of nature and culture" (*NBM*, 30). The paradox lies in how each form of critique makes the other impossible. Because nature and culture are hyperseparated, their contact is, on a certain level, forbidden as a pollution or corruption of their purity. At the same time, however, we continually confront hybrid phenomena over which a large degree of control is possible due only to the analysis of their parts into the pure spheres of nature and culture. Complex hybrids are thereby reduced to simple essences, which renders their existence as a hybrid irrelevant, since anything they make possible is explicable through their more basic constituent essences. Speaking of essences here, it is important to remember that we are considering both nature *and* society. The paradox becomes powerful due to the two-sided maneuver that it makes possible: nature grounds the social (society immanent to nature), but transcends it and remains unchanging (purified nature), while the social world remains free to transform itself (purified society), even as it displays the regularity of certain laws and dynamics (nature immanent to society) (37). The dual immanence and transcendence of both nature and society allows the modernist to occupy and address every possible position from which one might challenge it.

This paradox is made possible through a tacit "constitution" that Latour argues embodies the political project of modernity. The function of the constitution is to "render the work of mediation that assembles hybrids invisible, unthinkable, unrepresentable" and therefore "allow the expanded proliferation of the hybrids whose existence, whose very possibility, it denies" (*NBM*, 34; emphasis removed). Describing this situation in detail will lead us toward the heart of Latour's complaint about nature, for it is nature that is one of the main conceptual agents of modernist dualism. The concealment of the work of mediation that occurs through the hyperseparation of nature and culture is embodied in the first two guarantees of the constitution: (1) "It is not men who make Nature; Nature has always existed and has always already been there; we are only discovering its secrets"; and (2) "human beings, and only human beings, are the ones who construct society and freely determine their own destiny" (30). These two guarantees secure the transcendence of nature and society respectively, but the transcendence is guaranteed only by their mutual immanence. The problem faced is that if nature and society are taken as separate ontological spheres as the guarantees suggest, Nature will remain aloof and impenetrable since any knowledge that we attain must be "discovered" by some human means in a form that is socially intelligible; and Society will lack any firm foundation, which leaves it susceptible to various kinds of challenges to the social order. The solution to these difficulties is a practical one, wherein the laws of nature are discovered only through the *social* labor of scientists and technicians, and the society created through the freedom of humanity is not ephemeral and arbitrary, but, *like nature*, has its own durability and organization. The trouble with this solution is that by revealing the mediation of the social in our knowledge of nature and of the natural in justifying our social organizations, the solution effaces the transcendence that is necessary to assure the authority of the original guarantee. Thus, each sphere without the other remains unstable and another guarantee is necessary: (3) "There shall exist a complete separation between the natural world (constructed, nevertheless, by man) and the social world (sustained, nevertheless, by things); secondly, there shall exist a total separation between the work of hybrids and the work of purification" (31). Of course, this third guarantee is just the paradox of modernity. So we can see how the very conceptions of nature and society respectively lead to their hyperseparation because it is only through that hyperseparation that the two original conceptions can be made both consistent and stable. The hyperseparation has the result, however, of denying the social mediation of our knowledge of nature and the natural mediation of our social organizations and relations. For this reason, Latour argues that modernity demands the suppression of the very thing that it needs in order to propagate itself, the formation of hybrid networks of relations. The work of slowing down, then, is invaluable because it serves to bring to light these processes of mediation that are actively suppressed and obscured.

The modern constitution has one remaining aspect that should be mentioned: what Latour calls "the crossed-out God" (*NBM*, 32–35). Prior to modernity, the conception of God was marked by the same combination of transcendence and immanence. While there were many ways in which this combination was theologically conceived, in modernity God becomes "crossed-out" because neither society nor nature considered in a hyperseparated manner requires the existence of a god to guarantee its autonomy. He explains this conception of God directly as "the name given to a theory of action, mastery, and creation that served as the foundation of the modernist settlement" (*PH*, 267). Nature functions according to laws that require no god to explain their coherence, and Society is governed by a sovereign power (or democratic head-of-state) that can supply the laws for conduct that gods once supplied. At the same time, however, the place that God held within the metaphysical schema, namely, that of the source and guarantor of order, is retained in new forms such as the theological view that God is immanent within the hearts of human beings or the conception of a transcendent and ahistorical truth. In this way, the fourth guarantee assures a form of arbitration when nature and society conflict, but does so in such a way that places the deliberative processes of the arbitration outside the reach of humanity (*NBM*, 141).

The modern constitution is able to support itself so long as the number of illicit hybrids that it creates remains reasonable, but eventually the act of purification can no longer be justified without an explicit consideration of the practices of mediation or translation. Hybrid phenomena such as climate change surround us to such an extent that the pure categories of nature and culture no longer seem appropriate ways of carving up the world (*NBM*, 50).[4] A number of Latour's studies demonstrate this principle, and these hybrids of human and nonhumans (or simply of nonhumans) are what he calls "actants" (*PN*, 75).[5] What is important for this inquiry is that the basis for rejecting the modern way of dividing up the world is found in the epistemic and political limitations it experiences in accounting for hybrids. While the epistemic difficulty is obvious insofar as the practices of purification and translation result in a paradox when we seek to understand the world dualistically, the political limitations deserve mention. The strict separation of nature and culture, upon which the consistency of the constitution rests, results in the disempowerment of *both* humans and nonhumans. This point is worthy of note, since on most accounts in environmental ethics an active humanity dominates a passive nature. On Latour's account, everyone and everything loses something, which puts him in broad agreement with other philosophers such as David Abram and Val Plumwood: nonhumans are silenced by exiling them from the domains of meaning and agency, and humans are silenced by being relegated to the order of mute nature (e.g., women and indigenous peoples).[6] The silencing of the nonhumans leads not only to their domination via ignoring their contribution to agency, but also to stripping them of a history. Nonhumans, particularly nonliving things, are thought to exist in an ahistorical and unchanging manner. While the situation has improved for living beings since the spread of the theory of evolution, philosophies that continue to deny that nonliving things have history or agency in any relevant sense still predominate, contributing to their continued disenfranchisement. To combat this tendency, Latour will argue that "what Sartre said of humans—that their existence precedes their essence—has to be said of all actants" (*NBM*, 86). This notion will be a significant point of convergence between Latour and the phenomenological tradition as we proceed with the argument. On the human side, the problem is twofold: the aforementioned disenfranchisement of those groups that have been deemed more natural, as well as the establishment of the unquestionable authority designed to make the political process undemocratic (*NBM*, 8–9). If there is a nature to which one can appeal in order to justify one's political project, then one's opposition has little choice but to accede to one's political will. After all, one cannot challenge the constitution of the cosmos. In this way, even those included in the polis can still be disenfranchised by finding themselves on the wrong side of nature. Hence, Nature operates as a means by which social hierarchies are maintained, with examples including the enslavement of Africans, the project of colonization in general, the subordination of women, and the contemporary failure to extend certain civil rights to homosexuals.[7] The modern constitution therefore establishes relations of domination at both levels and can therefore be said to fail.

Asserting the failure of the constitution, however, does *not* mean that Latour believes it needs to be completely abandoned. Responding to the question of what is to be retained from the modern constitution despite its failure, Latour responds, "Everything, apart from exclusive confidence in the upper half of the Constitution, because this Constitution will need to be amended somewhat to include its lower half too" (*NBM*, 132). The lower half of the constitution, the work of translation that produces hybrids and thereby shores up the work of nature and society as ontological categories, can be brought into the constitution explicitly only by reformulating the relations of purification that represent the work of the upper half of the constitution. More simply, Latour seeks to eliminate the explicitly dualistic operation of purification and replace it with a consideration of the co-constitution of nature and society.[8] I will treat Latour's account of how establishing an ontological continuity between all beings reveals a different possibility for thinking about the relations between humans and nonhumans in the third section of this chapter. For now, we must first turn to his reasons for rejecting the contemporary attempts to liberate nature from its domination by resisting the project of modernity wholesale.

II. REJECTING *NATURPOLITIK*

As we have seen, the problem with nature in the modern conception of the world lies both in the untenable character of the dualism between Nature and Society as exhibited in the existence of hybrid phenomena and in the practical consequences of establishing a hierarchical order in both domains. While many aspects of the above description of our contemporary situation with respect to metaphysical dualism are not novel to Latour, it is on the point of how to respond to the critique that he separates himself most radically from those with whom he for the most part agrees. Where environmentalists, staunch humanists aside, have predominantly been concerned with asserting the rights of nature, its intrinsic value, criticizing the exercise of technological dominion over it, and so on, Latour sees the reassertion of nature as somewhat misguided. The best response to the problems posed by the modern constitution is not to reject the modern constitution and become "antimodern," but to transform it in a "nonmodern" way, that is, in such a way that the processes of mediation that form complex actants are avowed and examined (*NBM*, 47). In environmental circles, antimodernism takes many forms, but we might characterize it in general in terms of a return to nature in which nature is taken to be a thing possessing properties of its own. The enemy, on many accounts, is the mechanistic conception of nature and thus to it is opposed the vision of nature as an organism.[9] In other words, even as environmentalists may resist the *project* of modernity, many continue to define their critique in the very categories of modernity: they "have accepted their adversaries' playing field" (*NBM*, 48); and in doing so, "under the pretext of protecting nature, the ecology movements have also retained the conception of nature that makes their political struggle hopeless" (*PN*, 19). As we can see, the disagreement with the ecology movements is not necessarily with their political struggle, since many of Latour's political suggestions are consonant with their goals. How else to understand his lament that "nonhumans are born free and everywhere they are in chains" (*PH*, 172), and his subsequent claim that his analyses set the nonhumans free (266)?[10] Rather, his reservations stem from the attempt to use modernity's logic in order to resist modernity. Accepting modernity's way of articulating the world, he argues, eliminates the possibility of removing ourselves from the structures of power that serve to perpetuate the unhealthy relationship to the nonhuman world. In this way, the undertaking we face in removing ourselves from the power structures of modernity is to multiply the models of political action rather than resisting modernity with its own model of power (*PH*, 280).

Latour calls the strategy of using the modern conception of nature as a means to exert a political force in defense of nature "*Naturpolitik*." *Naturpolitik* is a form of political ecology, but one that fails to realize the radical project with which it is charged. For Latour, political ecology is a project that has *always already been undertaken*, regardless of whether we realize we have done so (*PN*, 1). As we have seen in *We Have Never Been Modern*, the modern world is defined by a specific dualistic relationship between politics and nature. The constitution conceals this political organization within the metaphysical and epistemological considerations that support the dualism. The organization becomes unquestionable because to question it is viewed as interjecting politics into the ontological realm, which is seen in turn as an absurdity. In this way, the political ecology of modernity is hidden from view. Challenging this hidden order is not as simple as proposing an alternative order, however. To do so would simply stealthily replace one organization with another, and we would therefore retain all the previously discussed problems concerning the undemocratic nature of that order.[11] Here is the trouble with *Naturpolitik*: as antimodernism, it remains ensconced in the moderns' hidden political order.

Antimodernism is, in Latour's view, a reactionary project (*NBM*, 9). *Naturpolitik* accepts the modern way of envisioning the world, even as it rejects the consequences of that vision in the name of resisting the domination of nature. This quixotic endeavor follows from a conviction that the things of value under the modern constitution (e.g., "localities, or spirit, or rationality, or the past, or universality, or liberty, or society, or God") are valid in the way modernity defines them (*NBM*, 47). In the same way that the moderns used the idea of nature as a means of maintaining a distance between humanity and the nonhuman world while simultaneously relying upon the presumed solidity of nature to buttress the social structure, practitioners of *Naturpolitik* are guilty of attempting to foreclose or short-circuit the work of politics in the name of the same values (*PN*, 30). Citing Baird Callicott as a prime example of this tendency, Latour indicates how attempts to save nature by reconceptualizing it in terms of an organic, holistic system continue to suffer from the same deficiency as modernity, unifying into an organized whole what can be unified only through political process (*PN*, 258n35). Because the theory of political ecology surreptitiously pursues an act of unification in order to provide a foundation for its "militant practice," modernity continues to define the antimodern theory of political ecology, disabling the radicalism of its political demands, namely, the political recognition of nonhumans. For this reason, Latour recommends, "from now on, whenever people talk to us about nature, whether to defend it, control it, protect it, or ignore it, we will know that they are thereby designating *the second house of a public life that they wish to paralyze*" (*PN*, 18). Opposing this paralysis requires political ecology to be practiced openly and not in the shadows.

In order to understand how the progressive concern for nature remains within the confines of a theoretical model designed to strip the nonhuman of its political force, we must delve more deeply into what is problematic about the concept itself and not merely the role it plays within the modern constitution. In other words, if we are to reconfigure the relations between the upper and lower halves of the modern constitution as Latour suggests, we need to know more about how nature functions within that economy. As stated above, "nature" is the problematic name for "the second house of a public life," the first of which is "society," the human domain. Nature paralyzes the relations because, as we have seen, it prematurely unifies all the nonhumans into a transcendent and, at a certain level, unchanging order.[12] Thus, the problem with nature is not with nature per se, but with the way the concept is utilized to preempt consideration both of the processes by means of which the world around us is unified and of the fact that *there may be more than one order*. Latour expresses the thought as follows: "The whole power of the term [nature] comes from the fact that it is always used in the singular, as 'nature in general.' When one appeals to the notion of nature, *the assemblage that it authorizes counts for infinitely more than the ontological quality of 'naturalness,' whose origin it would guarantee*. With nature, two birds are killed with one stone: a being is classified by its belonging to a certain domain of reality, and it is classified in a unified hierarchy extending from the largest being to the smallest" (*PN*, 28–29). So even as it challenges the dualistic conception of the humanity and nature, *Naturpolitik* retains this function of nature, that of ensuring that a unified, hierarchical order serves as the foundation for our normative views. Latour underscores this idea by saying that "it has always been the *unity* of nature that produces its entire political benefit, since only this assembling, this ordering, can serve as a direct rival to *the other form* of assembling, composing, unifying, the entirely traditional form that has always been called *politics*, in the singular" (29). "Naturalness" can remain such a strong normative concept precisely because it purports to lack the arbitrariness of the social order due to its transcendence. If its transcendence is denied, then the appeal to naturalness loses its normative force; it merely refers to *an* order, and one that could, in principle, be different. I am not speaking here of the commonplace idea that the world could have followed a different evolutionary path and appear very different from how it does today. Rather, the idea is that *nonhumans themselves have a history* and are engaged in a *process* of unification such that the way the world is organized can and does change, such as before and after Pasteur's "discovery" of microbes.[13] Rejecting *Naturpolitik*, then, requires us to grasp this historicity of things and reject the notion that there is *an* order of nature, any *one way* the world is supposed to be ordered: we must engage in *political* ecology in order to negotiate the constitution of the common world. In Latour's own words, "Politics has to get back to work *without* the transcendence of nature: here is the historical phenomenon that we are forcing ourselves to comprehend" (56).

When we cease to speak of Nature, new possibilities open up for how we could envision the work of politics. Since Nature is one of the main means by which the colonizing, oppressive practices toward non-Western people and nonhumans continue to be justified (even if only through the myth of progress), abandoning the concept of nature eliminates one of the main conceptual supports for those practices. In the place of Nature, we are left with "only the banality of multiple associations of humans and nonhumans waiting for their unity to be provided by work carried out by the collective, which has to be specified through the use of the resources, concepts, and institutions of all peoples who may be called upon to live in common on an earth that might become, through the long work of collection, the same earth for all" (*PN*, 46). Here Latour introduces some of his novel terms that are to replace the former ontological categories. The first thing to note is that the process of unification is not itself the problem, and, in fact, as I will argue is the case for Heidegger in the next chapter, is likely unavoidable. Rather, the concern is that unification happens in such a clandestine way that it establishes nature as transcendent, which is why he also indicates that his own unique contribution is bringing to light the processes of mediation that heighten the possibilities of relation that constitutes the real work of the moderns (*NBM*, 40–41). When that occurs, the political process is derailed. Removing nature from its status as an ontological category establishes an ontological continuity within which humans and nonhumans are able to commingle. This situation is what he calls above "the collective": through various processes (enacted by the sciences, politicians, and moralists, respectively), we are called to recognize various beings as existing and then reorganize the relations between already accepted beings in order to accommodate the new members of the collective. These processes of collection were formerly carried out dualistically such that only natural or social phenomena were said to really exist, which resulted in the problems already discussed. For this reason, "we shall have to avoid *both the notion of culture and the notion of nature*," in order to begin re-collecting the collective in a more democratic way (*PN*, 48). He warns us, however, to be cautious in how we understand what is going on here: "If we use the word 'collective' in the singular, it is thus not in order to signal the same type of unity as the one implied by the term 'nature,' and still less to designate a utopian 'reconciliation between man and nature'" (*PN*, 59). The collective, rather, is the result of an "experimental metaphysics," by means of which a kind of minimum set of requirements, a "metaphysical 'minimum wage,'" is instantiated in order to determine what does and does not gain entry into the collective (*PN*, 60–61). It is this more democratically organized collective that he calls the "common world." We will explore this vision in what follows, but in explaining Latour's positive views, we will also begin to engage with them somewhat more critically in order to ascertain whether and how a positive conception of nature might be retained despite these concerns about how the concept is normally deployed.

III. RE-COLLECTING THE PLURIVERSE INTO A COMMON WORLD

The situation in which we find ourselves is one where the tensions in the modern constitution, precisely because it denies the existence of the processes of mediation that form complex actants, have made the dualistic worldview implausible, but we have not yet formulated a coherent alternative set of concepts that would re-collect the world differently so as to avoid these tensions. The task of political ecology is to find a way to re-collect without relying on a unified and static conception of nature in order to do so. Latour succinctly summarizes the need to move away from nature in order to re-collect a differently organized collective: "If 'nature' is what makes it possible to recapitulate the hierarchy of beings in a single ordered series, political ecology is always manifested, in practice, by the destruction of the idea of nature" (*PN*, 25). The key here is that nature is being narrowly conceived as "a single, ordered series." By destroying that idea of nature, political ecology no longer has to consider the world as possessing an inherent and unalterable order and can set itself to disclosing different ways to conceive of and interact with the world around us. Hence, political ecology begins the work of governing a common world, the work of developing a politics without nature.

The first step in the process of developing a politics without nature is to clarify what exactly Latour means when he speaks about political ecology. To do so, it is important to understand that he is referring to a specific sense of politics. Rather than the traditional definition of politics that emerges as a foil for nature within the social world, the politics suitable to the nonmodern world will be identified with "the entire set of tasks *that allow the progressive composition of the common world*" (*PN*, 53, 18). This seemingly simple definition contains within it several important differences with the modern constitution. Most markedly, it consolidates within politics the work of the modern conception of both nature and culture. If politics is "what *gathers* everything together" in the collective, the resulting unification is *analogous* to the old nature's role of unifying, but with at least two important differences: the resulting unity (1) is a multiplicity of beings, not a substantial domain or itself a being; and (2) is something to be brought about, not an order established in advance. Since society, a culture, is usually the domain in which the order is brought about through a deliberative process, the unity is nature-like only in relation to a culture-like set of practices. Accordingly, Latour calls the collective that results from political unification a "nature-culture," that is, a way of dividing beings according to their properties, powers, and hierarchical relations while holding them together in terms of their relationships (*NBM*, 107). Ecology, as the science of the relationships within ecological systems, should also be heard in a slightly expanded sense as referring to studying these relationships in general. The usage is akin to the understanding of the term when someone speaks of, for example, "human ecology." Since there are many ways to convoke a collective (i.e., different practices, interpretations, and operations give rise to distinctly articulated wholes), there will always be multiple nature-cultures, each of which possesses a distinct ecology. For this reason, anthropology takes on a particular importance for Latour insofar as a certain way of practicing it reveals various ways in which human groups have sought to convoke different kinds of collectives from the modernist nature-culture.[14]

The relevant focus for political ecology, then, is *how* such gathering occurs and *by whom* it is enacted. For this reason, it behooves us to say a word first about *what* is being collected. For Latour, we always find ourselves within what he calls "the pluriverse," which he defines as "the multiplicity that no collective yet collects" (*PN*, 40).[15] In many ways, the pluriverse is akin to what Gilles Deleuze and Félix Guattari have called "the Earth," the completely deterritorialized plane of consistency.[16] It is a domain of forces, affects, and intensities not yet organized, not yet collected and unified, perhaps not yet even individuated. The pluriverse is what the sciences confront at the limits of their inquiries, where affects can be reified into substances and new objects of inquiry constructed. At a less abstract level, the pluriverse is preliminarily articulated into "propositions," by which Latour refers to attempts at individuating an actant from out of the pluriverse that has not yet been accepted into the collective. The sciences make propositions to the collective, as, for example, is occurring in the current debate over string theory. Physicists are divided between those who believe introducing quantum strings as actants in the collective would finally resolve the problem of unifying the four fundamental forces and those who resist introducing the strings for various reasons (e.g., lack of experimental verifiability). Most physicists continue the debate upon the background of a nature the structure of which string theory is meant to express in mathematical terms; in other words, the debate is about whether or not the theory accurately describes a being in nature. Another way to view the debate is whether or not individuating quantum strings as actants is actually a good way of articulating the various relations that occur at the quantum level. These relations perplex us, and the question is whether a phenomenon like quantum strings will allow us to integrate the perplexing relations by means of the activity of a substance. On this interpretation, a substance is not something other than its properties and powers, that is, a substance is something made rather than discovered; a substance is "what gathers together a multiplicity of agents into a stable and coherent whole" (*PH*, 151). Obviously, this gathering is object-oriented: one can go about such gathering only in ways that respect the phenomena observed. This idea will return and be discussed in detail in the coming chapters. In the case at hand, however, the agents to be unified are the forces of interaction, which are themselves former propositions that have been accepted into the collective. Quantum strings propose to bring together these agencies by explaining their existence through another agency. In this way, new propositions name new substances insofar as they "turn a new entity into what *lies beneath other entities*" (*PH*, 170). In other words, a substance is an *institution*, a way of concretely organizing the real world that confronts us into a coherent set of propositions. These institutions then serve as the foundation for the ascription of "naturalness," "which is very convenient for designating, on a routine basis, full-fledged members of the collective" (*PN*, 114). The work of the sciences is to determine what should and should not be admitted into the collective without resorting back to the metaphysics of nature. At the same time, however, we can see in Latour's use of the idea of being "natural" a space emerging from within which nature might remain a meaningful concept, though it will have a very different sense than it did within the modern constitution.

In order to see what this reformulated conception of nature might be like, we need to take a closer look at how this work of articulating propositions and collecting them into a common world remains a political project.[17] Recall that the complaint concerning the metaphysics of nature is that it settles the political questions surrounding the constitution of the collective in advance and without discussion. That being the case, Latour describes the appropriate response to this situation in what is a lengthy passage, but one that bears reproduction:

> It is understandable that people find it hard to give up the conveniences procured by such an arbitrage between the indisputable and the disputable [as bifurcating nature and society into separate ontological domains accomplishes]. The Constitution that we seek to draw up affirms, on the contrary, that the only way to compose a common world, and thus to escape later on from a multiplicity of interests and a plurality of beliefs, consists precisely in *not* dividing up at the outset and without due process what is common and what is private, what is objective and what is subjective. Whereas the moral question of the common good was separated from the physical and epistemological question of the common world, we maintain, on the contrary, that these questions must be brought together so that the question of the *good* common world, of the *best* of possible worlds, of the *cosmos*, can be raised again from scratch. (*PN*, 93)

By envisioning collectives as the result of a political procedure of instituting an order among beings, Latour envisions several key benefits that were not available to the modern constitution. If the collective is a process of unification, the common world is the result. The processes, however, are not value-neutral: both the decisions as to what will constitute the various articulations of the pluriverse and the decisions as to which beings count as a part of which categories are inherently *political* and therefore guided by values.[18] These values, whatever they turn out to be, are what help to shape and guide the common world toward being the *best* of possible worlds. Now, the Leibnizian language may make some uncomfortable as it hearkens to a teleological conception of both metaphysics and history, but there is a critical difference between the kinds of ideals that are guiding the two respective conceptions of the best possible world.[19] For Leibniz, that world already existed; it was our own, fabricated to be the best by an all-knowing, all-powerful, beneficent god. For Latour, the best of possible worlds is *to be made*; it is something to be pursued only by means of political laboring. As such, no one value or set of values is given in advance to guide the collecting, and any values we adopt can be adopted only provisionally and must be subject to continual renegotiation as new members enter the collective. In other words, there is no one way the best of possible worlds might be, and it can always be made otherwise. The conception of what makes the world the best possible world is the subject of debate.

Another possible misunderstanding of the project of instituting the best of possible worlds is that re-collecting the pluriverse into new visions of a common world is sufficient to improve upon the common world of the moderns. It is simply not the case that a political process will be a guaranteed success. Just because relationships are reorganized and conceived differently does not assure that a new collective is going to be less dominating or that it will lead to an even less desirable political situation than the one embodied in the collective it replaces.[20] As he puts the point in another context, political ecology does not promise peace, that is, a harmonious whole in which all beings exist without tension or conflict (*PN*, 219). There are, however, certain basic principles that Latour offers in order to guide the collection in a fair way: (1) the membership of the common world is always in the process of expanding; (2) the sciences come to be understood as a means of determining membership in the common world; and (3) new members of the common world are introduced to add to the richness of relations within the world rather than to short-circuit the political processes being negotiated between current members (*PN*, 38). While it would take us too far afield to discuss in detail his attempt to reinterpret Science in terms of the practice of the sciences, what is important to note is that they retain their position as the arbiters of what is and is not a substance acknowledged by the collective (i.e., a being), but they retain that position only by means of a reformulated sense of objectivity: the sciences are what allow silent nature to speak in human language. Latour, like the phenomenologists we will examine in the next three chapters, believes that beings themselves possess a sense, and that "thought is seized, modified, altered, possessed by nonhumans, who, in their turn, given this opportunity by the scientists' work, alter their trajectories, destinies, histories" (*PH*, 282). Part of the reason for the reformulation of objectivity lies in the fact that all description is normative (*PN*, 224), but the more pressing concern is that once beings have been removed from their dualistic metaphysical context in which natural things were thought to be inert and passive, we can begin to see them as having their own powers and form of agency. The result is a diminishment of the priority of individual agency,[21] but an enriched conception of the agency of actants, the assemblages of relations between substances. The task, then, in dealing with the struggles of coexisting with myriad assemblages of actants, each pursuing its own course, is to free nonhumans from a dualistic understanding of them as mere objects and recognize this agency of actants.

At this point it is appropriate to pause and compare this project to restore the agency of actants with that of Plumwood and others. If, as Latour argues, "objects are not innocent inhabitants of the world" (*PN*, 50), and we proceed to recognize a certain agency possessed by actants, a "thing-power" in the words of Jane Bennett, we might also be tempted to argue, as do Plumwood and others, for the autonomy of nature.[22] In a certain sense, doing so would be a mistake. Rather than arguing for the autonomy of these liberated beings, he instead believes we need to understand them *in relation* to us and to one another. The point of distinction here is a complex one. Latour takes up his position so as to avoid resequestering beings within a new nature-like dualizing collector, by which I mean interpreting all beings as passive, fixed, unitary, and having an unchanging essence. To assert the autonomy of a being in this context is to purify it of its relations, to consider it as if that essence would be the same if it did not have the relations that it has. This purification, of course, includes its relations to humanity, which is exactly what the modernist conception of nature is meant to do—isolate nature from its relations to humanity such that it can serve as a normative foundation for a variety of social projects. So it seems that arguing for the autonomy of nature is wrongheaded at two distinct levels: (1) considering nature as a thing seems to assume a preexisting order; and (2) regarding nature as autonomous fixes on the wrong target, insofar as describing the function of nature independent of humanity reinscribes the dualistic tendency that the concept was developed to overcome. The matter is not so simple, however, due to the fact that extending the concept of autonomy to nature was meant to distribute humanity's intentional powers to nonhumans along a continuum and establish the ontological continuity between all beings. The motives for the two positions are, then, quite similar. The question becomes whether the more progressive solution to dualistic thinking lies in reformulating the category of autonomy in terms of relationism or in moving in the other direction and saying that only complex actants have agency, and these actants always include nonhumans. If autonomy is reconceptualized in terms of relationism, the difference between these two positions might seem merely semantic, but the distinction between them remains meaningful nevertheless: continuing to speak of the kind of agency uncovered through the political redistribution of powers as "autonomous" still locates a particular being's agency within itself, that is, it configures responsibility in a very specific form that still assumes a causal theory of interaction. Contrarily, if the redistribution of agency moves us away from understanding a being's essence in terms of its possession of certain properties and toward a conception of essence defined in terms of relations, it can provide a stronger platform for reconceiving of nature outside the boundaries that it has historically inhabited and that Latour has correctly criticized.

As was related above, "God" stands for a theory of action that places a premium value upon ideals of mastery and models itself on a linear causality. If we are to develop a world without masters, this theory of action is the one that needs to be abandoned. Autonomy, as a term that explicitly references the metaphysical conception of the subject within modernity—and this is explicitly its appeal for Plumwood—will remain indebted to modernity's theory of action. In spite of this connotation, we might retain a place for autonomy *as long as it does not refer to agency*. Latour will frequently refer to the autonomy of facts, for example, which is what is accomplished in the institutionalization of propositions described above (e.g., *PH*, 102–3). Agency, however, remains a property of actants, and actants are never autonomous in any traditional sense.[23] They are concrete and distinguishable, but they are always assemblages that find themselves defined through their relations with others. Actants can be individuated in terms of how "they modify other actors through a series of trials that can be listed thanks to some experimental protocol" (*PN*, 75). In short, agents are known through their affects and effects, their collective powers and alliances.

While this discussion of autonomy may seem a digression, it provides what is the last piece of the puzzle regarding how we might begin to reconstruct a nonmodern conception of nature. We have already seen that any nonmodern nature must be organized politically rather than unified in advance, and for that reason nature will never be a being or a substance in the modern sense of those terms. We can now add to that a positive possibility: it may be the case that nature is an actant, an assemblage of bodies contingently related such that there is an order, always subject to revision and reconstitution, that emerges from the interrelationships between smaller constituent bodies. It is this conception of nature for which I will argue in the following two chapters and the normative consequences of which I will describe in the conclusion of the book.

IV. CONCLUSION

The stage is now set for us to examine how this conception of nature might be justified and of what kinds of powers and alliances, affects and effects it is capable. To do that, I propose examining phenomenological philosophers' attempts to cultivate a conception of nature that relies upon neither of the traditional alternatives of mechanism or teleology. This third way of conceiving of nature in some ways may seem very similar to traditional views of nature, but that is only because all conceptions of nature are addressing the same environing world. If, as Latour argues along with phenomenologists, the things impinge upon us and demand to be understood in particular ways, it should not surprise that this conception of nature accords in certain ways with the image utilized within contemporary scientific ecology. At the same time, however, if we distance ourselves from the *modernist interpretation* of ecology, that is, from the assumption that ecology is Science and therefore describes nature as it is, we will find that ecology still cannot serve as an unproblematic and unquestionable foundation for normative views because any order it reveals is contingent.

The way to such a renovated conception of nature through the marriage of Latour's model for the nonmodern world and phenomenological inquiry is, unfortunately, a complicated path indeed. After all, a consistent theme throughout Latour's writings is a denigration of the importance of phenomenology and of Heidegger's phenomenology in specific. What I propose is to examine Heidegger's phenomenology as an example of the continuity between certain aspects of Latour's ideas and phenomenology. In the process, we will have to assess what, exactly, Latour's objections are to the project of phenomenology and how, if necessary, they can be surmounted.

CHAPTER 2

SCIENCE, TECHNOLOGY AND THE CLOSURE OF NATURE

> There is then in all technical processes a meaning, not invented or made by us, which lays claim to what man does and leaves undone. We do not know the significance of the uncanny increasing dominance of atomic technology. *The meaning pervading technology hides itself*. But if we explicitly and continuously heed the fact that such hidden meaning touches us everywhere in the world of technology, we stand at once within the realm of that which hides itself from us, and hides itself just in approaching us. That which shows itself and at the same time withdraws is the essential trait of what we call the mystery. I call the comportment which enables us to keep open to the meaning hidden in technology, *openness to the mystery*.
>
> —Martin Heidegger, "Memorial Address"

While Heidegger is usually viewed as a severe critic of technology, there is another dimension to his thoughts about it that are expressed in the quote above: by remaining open to what we can learn from the meaning of our technological relations to the world, we learn something of the essential *mysteriousness* of being. There is no denying that Heidegger *is* a critic of the technological world, but the denunciations of him as a conservative romantic, regardless of their legitimacy, seem to miss this crucial connection between technology and what Heidegger finds most distinctive of human beings: the ability to question being as such. Engaging with technology brings to the fore humanity's ability to interpret the world *as* a world, to engage with beings in diverse ways, and, synthesizing these two, to change the ways we relate to beings *by* reinterpreting them. The criticism of technology follows from Heidegger's belief that technology surreptitiously assails our ability to relate in this way to beings by presenting beings as having one definitive meaning and truth. While Latour is much more optimistic regarding the potential impact of technology on our lives, it should be clear from this brief summary that there is a deep resonance between their views. Both resist the nonhumans' being reduced to their objectness, both want to conceive of different ways of collecting beings, and both are skeptical of there being any one, inherent order to the world that is not mediated by various kinds of relations.

In other ways, however, Heidegger is Latour's philosophical nemesis, with the latter directing numerous attacks against the former, ranging from the in-depth to the inane. While this has led some to believe that Latour is being somewhat disingenuous in his treatment of Heidegger's thought, I prefer to take Latour at face value, that he believes he and Heidegger have legitimate philosophical differences and that his own views concerning science and technology are to be preferred over Heidegger's.[1] Though we will delve into these views more deeply shortly, we might summarize Latour's views about Heidegger by saying he finds that Heidegger understands technology in too limited a fashion, which leads him in a distinctly antimodern direction. Though I do agree that there are both dispositional and meaningful conceptual differences between the two thinkers, I am also inclined to agree with Søren Riis that there is more similarity between their views than Latour acknowledges (though not exactly in the way Riis imagines).[2] One of these points of connection is Heidegger's views on nature, which is why we will have to pause briefly in order to take stock of the relationship between Heidegger and Latour as a preliminary step in our analysis of nature.[3] My intention here is not to defend a definitive position regarding the ultimate compatibility of their philosophies as a whole, but rather to examine this specific point of connection between them in order to see if there is any room for the intuitions that guide Heidegger's philosophy in our inchoate sense of what a nonmodern conception of nature is like.

The main point of this chapter, however, is to begin the process of constructing a nonmodern conception of nature by laying out Heidegger's own criticisms of modernity and the role of nature within it. Heidegger, like Latour, was rather devoted to the displacement of the primacy of nature within philosophy. Though Heidegger tends to refer to "earth" when he speaks about the environment and what he takes to be the human project of mastery toward it, his goal is not, strictly speaking, to disallow thinking with respect to nature. Rather, he seeks to redefine what we speak about when we speak of nature in such a way that nature is no longer taken to be either a totalizing object or a system of objects in a preexisting space and time. In Heidegger's view, if nature is taken as a preorganized system, a being, or a substantial domain (i.e., the traditional conception of matter), it becomes closed and subordinated to an ordering principle.[4] In rejecting the systematization of nature, Heidegger seeks to *open nature up* in order to cease understanding the surrounding world as a totalized, ordered realm. Part of this attempt, however, involves a reevaluation of the notion of nature itself. Consequently, we must proceed with caution in endeavoring to speak of nature while at the same time not collapsing back into the sense of nature that Heidegger is attempting to dislodge.

In order to understand why Heidegger wants to cease speaking of nature in its ordinary sense, we will begin with a brief discussion of his "history of being," focusing specific attention on the foundational role played by the Greek notion of *physis* therein. Heidegger's account of the history of being relies upon a distinction that he calls "ontological difference," which we can understand as the distinction between being as such and the being of beings. To speak quite generally for a moment in order to elucidate the distinction, we can understand what Heidegger means by the "being of beings" as the general interpretation of what makes each being what it is; and by "being as such," he intends the general way in which human beings project open a world in which such interpretations are possible. This distinction will become clearer as we proceed, but it is important to introduce it now given that the entire history of being is predicated upon it. For Heidegger, the being of beings is fixed for the first time in Greek thought with the word "*physis*." *Physis*, of course, is usually translated as "nature," but it is precisely this translation that Heidegger resists, preferring instead to translate the term as "emergence" (*Aufgehen*), an opening up or blossoming. Heidegger's complaint is that when *physis* is translated from the Greek into Latin as *natura*, there is a limitation of the word's original force due to a forgetting of the original openness to the mystery of being that called for that interpretation of beings. Obviously, we will have to explore exactly what this mystery of being is in greater detail, but to preview what is to come in the second section of this chapter, the story Heidegger will tell about the history of beings will go as follows. In Greek thought there is no disambiguation between what he calls the grounding question of metaphysics—the question of being as such—and the guiding question of metaphysics—the question of what is the being of beings. For this reason, as each generation of philosophers responds to the guiding question in a different way, we grow further and further away from reflecting upon the grounding question. In other words, as the interpretation of the being of beings changes over time—for example, from *physis* as an inner movement of becoming, to nature as a substantial domain of matter—it becomes less and less clear that it is possible to *think* about being as such. Thinking, here, refers to asking the grounding question explicitly and questioning the very way in which human beings project open a world, which is different from saying a subject represents objects to itself. Identifying the ways in which Heidegger distances himself from the more subjectivisitic understandings of phenomenology is the subject of the first section of this chapter.

In addressing our contemporary philosophical and environmental situation through these ideas, we must resist the temptation to which some have succumbed, namely to argue that a *return* to this original sense of *physis* is necessary.[5] In addition to the reasons internal to Heidegger's philosophy for not doing so that I will explain, too much has happened philosophically and historically for such a return to be possible. I suggest a better way to understand Heidegger's efforts to recover what Greek thinkers such as Aristotle meant by *physis* is to view it as a reminder that *we can always think differently about being* and that *there are ethical consequences to different ways of thinking.*

Once we have come to an understanding of what Heidegger takes to be the original meaning of *physis* and the role that it plays in Greek thinking, we will turn our critical attention to the relation between these views on nature and his critique of technology. These views more than any others are what draw Latour's skepticism about the Heideggerian project, and so in many ways the potential allegiance between them will coalesce or fracture around whether their dispute is an *ontological* one that would prevent their working together, or one that is a meaningful difference *with respect to technology alone*. Latour's view is that the difference at the surface level reveals a deeper difference in their philosophies, but I will argue that any existing differences are in principle surmountable and that their disagreement is, to a certain extent, one between allies. With this in mind, let us begin.

I. PHENOMENOLOGY AND SUBJECTIVISM

In many ways, Latour's attempts to distance himself from phenomenology rest upon a well-established criticism: that phenomenology remains too subjectivisitic to be useful in understanding the world in which we live. In *Pandora's Hope*, he identifies the point of dispute as being phenomenology's reliance upon lived-experience as the foundation for its analysis and the way it relegates the sciences to a world that is "entirely cold, absolutely inhuman" (*PH*, 9). In opposing the world that the sciences disclose to the world that is disclosed in lived-experience, phenomenology ends up interpreting the work of the sciences as a particularly sterile mode of lived-experience. Husserl's views concerning the derivative nature of the naturalistic attitude from the personalistic attitude in *Ideas II* can serve as a convenient, if a bit unfair, touchstone on this point. While I find this criticism to be appropriate to a very narrow way of understanding the phenomenological project, an interpretation Quentin Meillassoux has recently named "correlationism," there has also been a strain of phenomenology that has sought to distance itself from the foundation of lived-experience, even as it retains experience as its object of inquiry and point of contact with the world.[6] The reason I have chosen to focus on Heidegger and Merleau-Ponty to develop the phenomenological contribution to an environmental ethic is that both of them exhibit a preference for the kind of phenomenology that views the philosophy of lived-experience to be somewhat limited by its traditional humanism. With that said, though, it is important to understand what the problems might be for the more traditional view and who believes the more problematic position among contemporary commentators before proceeding.

Latour situates phenomenology among the philosophies of the twentieth century that strove to secure the modern constitution by using its key terms to explain the existence of the problematic hybrid phenomena discussed in the first chapter. Picking up where the Hegelian/Marxist dialectic left off with the perpetual alternation between the subjective/social position on the one side and the objective/natural position on the other, phenomenology is accused of absorbing both positions by means of the theory of intentionality. Phenomenology "jettisoned the two poles of pure consciousness and pure object and spread itself, literally, over the middle, in an attempt to cover the now gaping hole that it sensed it could no longer absorb" (*NBM*, 57–58). As is well known, the first of Husserl's phenomenological reductions is to "bracket" the natural attitude, that is, to suspend our ordinary beliefs in the existence and nature of things in order to understand the structures that allow the phenomenon to appear as it does to consciousness.[7] Thus, there is no pure object because objects are always intentional objects, but there is equally no pure subject because consciousness requires an intentional object. The trouble with this view, for Latour, is that even as it strives to remove itself from the dualistic modernist practice of purification, it leaves the pure poles in place due to the way the theory of intentionality merely establishes a connection between subject and object, an insurmountable tension, rather than focusing on the practices of mediation that belie the validity of the poles themselves (*NBM*, 58). This way of regarding phenomenology wherein intentionality is an *act of consciousness* leads him to criticize it in the following way: "Phenomenology deals only with the world-for-a-human-consciousness. It will teach us a lot about how we never distance ourselves from what we see, how we never gaze at a distant spectacle, how we are always immersed in the world's rich and lived texture, but, alas, this knowledge will be of no use in accounting for how things really are, since we will never be able to escape from the narrow focus of human intentionality" (*PH*, 9). So the accusation is that phenomenology's reliance upon lived-experience traps us within our consciousnesses such that we never see "how things really are."

In many ways, this picture of phenomenology is warranted, if not within the primary texts then within those practitioners who would continue to insist on the primacy of subjective phenomenology over its more ontologically oriented counterparts. While any phenomenologist worth her salt will immediately criticize the notion that there is a way "things really are" independent of intentional relations, a notion that sounds close to Harman's autonomous objects, by pointing out that there is no way to experience something independently of intentional relations, there is a bit of truth to the criticism and it is important for us to understand what that truth is. *If* we unproblematically take phenomenology's point of departure to be lived-experience, then objects will always be constituted by a consciousness. As we will see, Heidegger criticized this way of understanding phenomenology long ago. But we could also understand Harman and Latour's point in this way: if intentionality is a specifically human property, that is, if humans are the only things that are thought to have *experience* in any meaningful sense, then the resulting relationship with the world will be inherently anthropomorphized. In this light, phenomenology very well may lay claim to being a relational ontology, but in reality it remains a veiled form of humanism.

There are several ways to respond to this criticism of the tradition. One would be to insist upon the theory of intentionality's ability to be naturalized in one way or another.[8] Another is to redistribute the presumed human properties among all beings. This strategy also has various forms, several of which have already been mentioned—Plumwood's attempt to attribute autonomy to nature, although she would likely not identify as a phenomenologist, or Abram's naturalistic animism both fall into this latter category. A third attempt to defend phenomenology, and this is the one we will explore in what follows, is to place it upon the foundation of a nonhumanistic relational ontology. Heidegger's philosophy lays the groundwork for this interpretation of the phenomenological project, but continues to remain humanist in the sense relevant above: the only ones with an access to being are human beings. It is not until the discussion of Merleau-Ponty's ontology of "flesh" in the final chapter that this project will be complete. The point of de-subjectifying phenomenology is to overcome the subject-object distinction, not by moving toward either pole or by creating an unalleviated tension between them *but by showing that everything that exists, exists as a nexus of relations.*

If this locution sounds familiar, it is because we already encountered it in the previous chapter when we examined Latour's view that every object is an event. This observation that objects are events is where I see the concerns of Latour and the phenomenological tradition aligning. In this vein, Lynnette Khong has pointed out that Heidegger and Latour agree on at least two points: (1) current misconceptions about the nature of technology and technological practice are based within a continued commitment to the subject-object distinction; and (2) alleviating the problems caused by this misunderstanding requires articulating an alternative form of relation to the subject-object distinction.[9] Though I agree with Khong broadly, this is a point of philosophy concerning which Latour can be rather slippery. For example, in some places Latour seems to be mocking those who would strive to "overcome" the distinction between subject and object because he views it as an inescapable conceptual net; it was developed so as not to be overcome (e.g., *PH*, 9). In other places, and this is also the more prevalent trend, he expresses a desire "*to avoid using* the subject-object distinction *at all* in order to talk about the folding of humans and nonhumans" (*PH*, 194). The seeming inconsistency is only on the surface, however. The reason he believes the subject-object distinction cannot be *overcome* is that *it never existed in the first place.* Of course, Husserl is also trying to establish that the distinction is derivative from a deeper level of lived-experience, but the point here is that one cannot *begin* from a position of indivision of subject and object *within the subject.* As long as that remains the starting point, transcendental philosophy will remain subject to the critique of subjectivism, which is why Latour observes that "all discussions on the abandonment of the distinction between object and subject all fall flat" (*PN*, 263n23). One must begin, rather, from an understanding of beings as events and then proceed to understand the kind of event at hand.

Justifying these last claims and showing their effects for an environmentally concerned philosophy will be the subject of the remainder of this book. If this connection between Latour and the specific form of phenomenology we are seeking can be established, neither of the views will remain unchanged. For Latour, there may be ways to think about a world without masters other than a bureaucratic structure governing over the constitution of the common world. For phenomenology, there may be a richer engagement with the world of things that would allow practitioners to contribute in a more concrete way to discussions of environmental concerns.[10] In order to make progress toward these positions, we will now turn our attention to Heidegger's philosophy to see in what ways he puts us upon this path and in what ways his philosophy might leave us with a further road to travel.

II. THE TWOFOLD ESSENCE OF *PHYSIS* AT THE INCEPTION OF PHILOSOPHY

For Heidegger, every genuine philosophical question arises out of the deeper question of "Who are we?"[11] When one asks a philosophical question, for example, our own question of how best to think of nature, there are certain presuppositions that must be in place in order to provide a response to the question. Specifically, the identity of the "we" is determined in advance such that the questioner is never put into question by the philosophical question. If this predetermination of the "we" is removed, however, the philosophical question is set into its proper place: a thoroughgoing inquiry into both the subject matter of the question and the being for whom the question is a question-worthy matter. In other words, unless the questions we ask call into question both the basic concepts under investigation and the meaning of humanity along with them, the responses to philosophical questions are all, in a certain sense, already mapped out in advance (*BP*, 53/38). The assumed determinations of language *organize in advance* any possible response to the question. If he is correct, then the first step along the path to any truly philosophical response to a philosophical question will be to question the most basic and foundational concepts within metaphysics in order to show that they are as indeterminate as the matter of the other questions.

This interrogation of the basic words of metaphysics is the motivation behind returning to the Greek word *physis* in order to see the ways in which this word and its history have determined in advance the history of philosophy. Heidegger's motivation is less to reject the word "nature" than to reject a specific way of comporting oneself toward the world that the word has come to represent. In *Introduction to Metaphysics*, Heidegger describes what he sees as the "naming force" of the Greek word *physis*, out of which our current word "nature" is but a hollow echo.[12] The motivation for such resuscitation, as we will see, is *not* a simple return to the Greek notions that Heidegger contends are thought on the basis of an experience of being. Rather, by returning to the original philosophical words, returning to the foundations from which all later philosophy is an unfolding, the hope is to accomplish two distinct goals. The first is to bring to the fore the specific questions to which humanity found itself in need of a response within the Greek world. In this way, because the Greeks were responding to the call of being, we, too, will learn to listen once more to this call and perhaps respond to it in another way. Following from the first, the second goal is to reorganize these relations in a nondominating manner, to establish a nondominating way of being. The specific claim is that, on the basis of the Greek concept of the being of beings as *physis*, a path was opened up that ultimately led to the technological domination of both humanity and the earth. In order to turn away from this technological domination, another pathway must be opened, one that remains as yet indistinct.

What is at stake, then, in the displacement of "nature" as the word for the being of beings is the relationship between humanity and the earth. "Nature," Heidegger says, "has become the fundamental word that designates essential relations that Western humanity has to beings, both to itself and to other beings."[13] No matter what content comes to fill in the word, the concept of "nature" has taken on the role of determining what a being is (the nature of X) and what possibilities belong to it (what is in its nature). It is no mistake that this use of nature is tied, in our contemporary discourse, to the movement of "naturalism," which is, broadly speaking, the attempt to decide what is real on the basis of a discovery of a physical, in this case material, cause for the phenomena. Naturalism determines what is possible on the basis of physical laws, that is, laws of nature. By re-placing "nature" within philosophical discourse, Heidegger is able to bring to the fore the ways in which the conception of nature employed in naturalism is *unnatural* when viewed in terms of the sense that "nature" and "physics" had for the ancient Greeks. Questioning the meaning of naturalism does not necessitate falling into "postmodernist deconstructionism" or a "social construction of nature," as a certain strand of environmental thought has rather poorly come to be called, but rather to acknowledge the myriad ways in which one's views concerning what nature is and how nature functions will demarcate the kinds of relations that are possible toward it.[14] In this sense, the original "nature" of philosophy, *physis*, established particular behaviors and relations as those proper to beings. Heidegger's claim is that many of the contemporary problems that face humanity, both social and environmental, have their sources in the maintenance of "nature" as the being of beings and the concomitant forgetting of the original sense of *physis* that made it possible for us to call the being of beings "nature" at all.

This idea is a delicate one, and one that is made more difficult by my current insistence upon maintaining the word "nature" in the face of Heidegger's refusal to do so. So, for the moment, let us put the word aside and begin, as Heidegger would have wanted, with the Greek sense of *physis*. One of Heidegger's most direct accounts of the sense of *physis* comes in *Introduction to Metaphysics*, but his richest and most in-depth analysis comes in "On the Essence and Concept of *Physis* in Aristotle's *Physics* B, I." We will begin, then, with *Introduction to Metaphysics* in order to provide a framework within which we might understand Heidegger's thinking on *physis* that the reading of the *Physics* will later fill in. In both texts, what is at issue is the meaning or truth of metaphysics. The hope is that by showing how metaphysics is programmatic on the basis of a given conception of physics, there will be the possibility of another mode of thinking that will not be beholden to this conception of physics. Part of what it means to step outside metaphysics, and by implication the traditional understanding of nature, is to propose a way of thinking that is not based on the being of beings defined as *physis*.

This contestation of metaphysical authority, of course, is quite a task, and one that must begin with the elaboration of the sense of physics that has dictated the history of metaphysics. The modern conception of the physical is derived from a series of transformations in the sense of *physis* from the "emerging, abiding sway [*aufgehend verweilenden Walten*]" (*EM*, 17/16) into that which is most stable, *Bestand*: the persistent, the constant, stock, or, as it is usually translated, standing-reserve.[15] At the inception of philosophy, *physis* had the sense of a coming into being or generation ([*Entstehen*]; *EM*, 17/16), but a coming into being of a specific character. Though merely asserted at this point, this conclusion is the result of a prolonged meditation on what Aristotle might mean by describing *physis* as a particular kind of movement, a kind of movement, moreover, that is the essence of all beings. Before entering into Heidegger's discussion of Aristotle's concept of *physis*, however, we must first look to how *physis* was originally deployed in Greek thought for the reason that Heidegger finds in Aristotle only an "echo of the great inception of Greek philosophy, the first inception of Western philosophy" (WBP, 298/229).

At the beginning of the Greek inception of philosophy we find thinkers like Heraclitus and Parmenides, and it is to these thinkers that Heidegger turns in his attempt to determine the originary sense of *physis* as the being of beings. *Physis*, as the emerging, abiding sway, names the manner in which a being comes to manifest (*erscheinen*) itself as the being that it is and thus, because the Greeks did not yet make the distinction between being itself and the being of beings, names being itself. This leads to the inception of metaphysics insofar as the Greeks ask what Heidegger calls the "guiding question"—What is a being? or What is the being of a given being?—as opposed to the "grounding question"—How does being essence (*west*)?[16] or What is the truth of being? (e.g., *BP*, 6–7/5).[17] The difference here is that in the case of the guiding question, one is asking how a being manifests, while in the grounding question the inquiry goes deeper in order to ask what structures are necessary for such a manifestation to be possible. While this may sound like a Kantian endeavor, it is important to note that the grounding question is not asking after the conditions of possibility of *experience*. Rather, Heidegger is asking about the conditions for manifestation in general. This distinction is an important one given the above criticism of phenomenology from Latour: we can already see Heidegger worrying about a more general interpretation of phenomena. So, in responding to the guiding question of metaphysics by identifying *physis* as the being of being, Greek thinking isolates a certain kind of manifesting, the emerging and abiding by means of which beings first become and remain observable, as being itself (*EM*, 17/15). As such, *physis* must also be connected to the Greek conception of "truth," *aletheia*. If *physis* is a coming into presence, it is a way of manifesting, but as a way of manifesting, *physis* is also unconcealment (*Unverborgenheit*), which is Heidegger's translation of *aletheia*. Unconcealment, however, is possible only on the basis of a world, which means, in terms of *Being and Time*, the system of relations within which individual beings appear as meaningful.[18] More specifically, Heidegger defines being-in-the-world as "the unthematic, circumspect *absorption* [*Aufgehen*] in the references constitutive for the handiness of the totality of useful things" (*SZ*, 76/71; emphasis added). While "absorption" is a fine translation of *Aufgehen* in this context, we should not fail to hear the connections that Heidegger is forging between the notions of world, *Aufgehen*, and manifesting in a nonrepresentational manner (the emergence is unthematic).

Turning back to *Introduction to Metaphysics*, we find that Heidegger explicitly makes these associations with the help of Heraclitus's fragment 53. In this fragment, Heraclitus is asserting that *polemos* is what allows for the differentiation of beings from one another in their respective manifesting, and allows for the establishment of "status and rank" such that "some appear as gods, others as human beings" (*EM*, 66/65).[19] Heidegger translates *polemos* as "*Auseinandersetzung*," another of his technical terms, which etymologically can be broken down into its constituent parts to say "placing one another outside of each other" or "setting apart," but is usually translated as "confrontation" or some variant thereof. Here, Heidegger defines the word in terms of a "strife": at the heart of manifestation, at the very center of the unconcealment of beings that is *physis*, there lies a strife that opens up, the creation of "clefts, intervals, distances, and joints" (*EM*, 66/65), that is, *articulations* within being. These articulations are encountered by means of a *polemos* between the world that finds meaning and status among beings and the beings that resist being projected open in this way. World preexists singular beings, certainly, but this does not mean that world is itself either a being or something independent of beings. World is held *open* in the confrontation, in the differentiation of beings; and, without being held open, it will collapse once again. Authentic history concerns not events in the world, but rather the transitions in articulations of beings, the changes in the interpretation of beings within a world.

We said that *physis*, though, is the manner in which beings come to presence (*anwesen*), that is, manifest. If this is so, then *polemos* is possible only through *physis*; beings can be articulated only insofar as they manifest. Thus, when the world is held open in strife, a task that Heidegger assigns to "creators, poets, thinkers, and statesmen,"[20] what holds the world open is *physis*, "coming to a stand in what comes to presence [*kommen . . . im Anwesenden zum Stand*]" (*EM*, 66/65). Since *polemos* is that which holds the world open, strife is not destructive. In fact, in bringing *physis* to a stand, *polemos* is that which allows a being to be the being that it is. In the strife, diverse manifestings are gathered together (*sammeln*) so as to present something constant (*ständig*), an individual being. This essential gatheredness Heraclitus calls *logos*, and because it is the gathering that brings the various emerging manifestations (*physis*) into an unconcealed (*aletheia*) constancy, Heidegger is going to assert, here at the inception of Western philosophy, an essential relation between *physis*, *aletheia*, and *logos*. Thus, in the struggle between the gathering and the gathered-together emergent manifestations within unconcealment, the stage is set for the meaning of being to become that of constant presence (*beständige Anwesenheit*): reality is defined in terms of what persists.

With the clarification of the Greek conception of being as constant presence through the essential relation of *physis*, *logos*, and *aletheia*, we are nearly prepared to see how our own contemporary conceptions of nature have arisen out of these ideas. Before that is possible, however, we must first make more precise the relation between these three terms so as to understand more fully their import within the larger aim of this book of establishing a conception of nature that is neither mechanistic nor teleological. In order to do so, we must turn away from the inception's beginning and toward its end in Aristotle's *Physics*.

Heidegger's perhaps infamous claim with respect to this text of Aristotle's is that it is "the hidden, and therefore never adequately studied, foundational book of Western philosophy" (WBP, /185). He argues this point, as we will see more clearly in the next section, precisely because of the appropriation of Aristotle in the Middle Ages and the subsequent forgetting of *physis*'s original sense that is marked in the word "nature." In Aristotle, Heidegger finds two distinct senses of *physis* that are in fact unified in a "simple essence" (296/228). On the one hand, there is *physis* as "the origin and ordering of the movedness of what moves of and by itself" (296/228). In his translation, Heidegger is attempting to show us how Aristotle, along with earlier thinkers, does not determine *physis* as a "cause" for a being's movement, but rather *physis* is both a source of movement and that which guides the movement along a certain path. Defining *physis* as a kind of movement, of course, does not mean that every physical being moves from place to place. Rather, the movement of *physis* is found in changes in coming to presence: for Aristotle, *physis* is still an emerging (247/191). As above, however, where the emergence is what brings a being to a stand, that is, provides the being with a certain constancy of its manifesting, Aristotle will call *physis* the being of beings, explicitly identifying it with *ousia*. Again Heidegger wants to resist the traditional translations of this word, finding its original sense in "what lies present" as opposed to "substance" or "essence" (258/199). The traditional translations serve to verify our contemporary commonsense understanding of nature as the substantial domain of matter. If *physis* is *ousia* in this other sense, the identification affirms the earlier findings that *physis* is a manner of presencing. In the current case, Aristotle goes even further insofar as he actually describes the manner in which *physis* makes a being present. As such, the connection to *aletheia* is also maintained insofar as the being of beings (*ousia*) is an emergence of a being into a constant manifestation.

In asking how *physis* itself appears as a movement of emerging internal to each being that makes it what it is, that is, how *physis* is *ousia*, Heidegger discovers the second sense of *physis*, as *morphe*, which leads to associations with both *genesis* and *kinesis* (WBP, 296/228). We must pause here to see how these terms are in fact related, and in unraveling this conceptual chain also come to see how this further determination of *physis* completes its forgotten essence.

The entry point into seeing these connections is Aristotle's distinction between *morphe* and *hyle*. Again, rather than relying upon the traditional translations of these two terms, "form" and "matter," respectively, Heidegger attempts to redefine them in terms of their semantic function within the Greek idiom. Rather than translating *hyle* as "matter," Heidegger takes up Aristotle's definition of *hyle* as "*to dynamei*," that which has an appropriateness for X. But for Aristotle, *morphe* and *hyle* act together in order to produce a being.[21] Taking advantage of the etymology of the German word for "to produce," *Herstellen*, which breaks down into "to set forth," Heidegger associates *hyle* with what is set forth and *morphe* with what sets forth. This shift in understanding results in *hyle* being translated as "the appropriate orderable" and *morphe* as "enframing in appearance [*Gestellung in das Aussehen*]" (WBP, 279/214, 274/211). In *Introduction to Metaphysics*, this appearing in terms of a look (*Aussehen*) is shown to be part of what it means for *physis* to be a manifesting (*Erscheinung*), but the appearing is a look that opens up the possibility for the concealment of what a being is (*EM*, 111/110). Aristotle confirms this interpretation when he claims that *morphe* "is *physis* to a greater degree than is *hyle*" because *morphe*, as a placing into appearance, is a movement of manifestation. *Morphe* is that which brings *hyle* to presence; it is the shape in which *hyle* takes on appearance. Hence, both of them constitute the essence of *physis*; they are not separable because without *hyle* there is nothing to enframe, and without *morphe* there is no order to *hyle*. All beings whose essence is *physis*, that is *ousia*, are constituted by a particular relation of *hyle* and *morphe*.

But what kind of movement is *morphe*, exactly? On this matter, the accounts in "On the Essence and Concept of *Physis* in Aristotle's *Physics* B, I" and *Introduction to Metaphysics* are in accord: for Aristotle, the movement of enframing in appearance, *morphe*, has the character of *entelecheia* (*EM*, 64–65/63; WBP, 280/215). If the *Physics* is the hidden book that determines the history of Western philosophy, Heidegger indicates that *entelecheia* is the central concept thereof, the "fundamental word [*Grundwort*]" of Aristotle's thinking and the "fundamental concept [*Grundbegriff*]" of metaphysics.

Entelecheia is defined in terms of a being's being held within certain limits, a movement oriented toward coming to a rest. Heidegger finds this movement in Aristotle's thought, "where rest [*Ruhe*] does not mean the breaking off and cessation of movement [*Bewegung*], but rather where movedness [*Bewegheit*] is gathered in *holding still* and where this ingathering, far from excluding movedness, includes and for the first time elucidates it" (WBP, 282/217). This moving toward a holding still is a movement toward an end (*Ende*, *telos*), where *telos* is not taken to be something outside the original movement, a goal (*Zeil*), as it is for Hegel; or a purpose (*Zweck*), as it was for Kant. Rather, *telos* as an end is the completion or fulfillment (*Vollendung*) of the movement of *physis* that does not close up that movement (*EM*, 65/63). The distinction between Heidegger's interpretation of *telos* as an end rather than as a goal or purpose is that the idea of both goal and purpose demand the stoppage of movement. Presumably, in differentiating this position from Hegel's and Kant's, Heidegger must also hold that the movement of *entelecheia* is something distinct from the circular or axial movement of Hegel's infinity.[22] In order to do so, Heidegger clarifies what it means to say that *telos* is both an end and a movement at the same time, or more precisely a rest that is a movement, by making the connection to gathering: *telos* is "the end where the movement of seeing first *gathers itself up* and essentially is movedness" (WBP, 282/217). Put this way, this formulation should remind us of his interpretation of Heraclitus, within which *polemos* was said to be the origin of the differentiation of beings within a world. As an end, *telos* is the result of the *polemos*, what is brought into being, but since *polemos* is never brought to an end, the *telos* is constantly holding itself within being through a renewed movement.

The movement of *entelecheia*, then, arrives at an end, a *telos*, that continually renews itself. But this arriving at an end, we said, is also the bringing of a being into being, generation (*Entstehung*). Through the movement of *entelecheia*, a being has been produced, but, in the case under discussion, this production of a being is literally a genesis (*genesis*) and not a production in any technical sense. Genesis is a movement toward the *telos*, which completes the dual sense of *physis* as *morphe* and *hyle*: *physis*, as *ousia*, as the movement of making present, is accomplished through an enframing in appearance by a genetic movement toward an end. Because *morphe* is the continual gathering into a *telos*, we may call it *genesis*. In the earlier interpretation of Heraclitus, however, this gathering into an end was achieved through *logos*, and we should not fail to see *logos* at work here as well. *Morphe* enframes a being in appearance, but, as a being, it must be gathered there in accordance with the originary strife or *polemos*. This gathering was said to be *logos*. But we also said at that point that the being thus brought to a stand in presence is unconcealed (*aletheia*) as a certain being. That is to say, it presents itself *as* something. What is unconcealed through *logos* is the *eidos*. As Heidegger expresses it: "*Eidos* means the appearance of a thing or a being in general, but appearance in the sense of the aspect, the 'look,' the view, *idea*, which it offers and can offer only because the being has put forth into this appearance and, standing in it, is present of and by itself—in a word, *is*" (WBP, 273/210). If *eidos* is what makes a being what it is, and it stands in unconcealment through *logos*, then *logos* unites *physis* and *eidos* as *ousia*. Expressed differently, if the movement of manifestation (*physis*) occurs by means of a gathering (*logos*) that makes the elements gathered together different from one another (*EM*, 132/131), and the result of the gathering is the appearance itself (*eidos*), then the *eidos* is brought into unconcealment (*aletheia*) only through *physis*.

In reviewing these foundational relations between *physis*, *aletheia*, *eidos*, and *logos*, we have finally arrived at a position from which we are capable of seeing what Heidegger believes has been forgotten in the essence of *physis* and how this forgetting, exemplified in the translation of *physis* as "nature," has led us to the historical position wherein the very integrity of the earth is threatened. The last two lines of *Physics* B, I, that Heidegger examines indicate this forgetfulness. Heidegger translates them as "And so this, the enframing in appearance, is *physis*"; and "However, the enframing in appearance—and thus that means *physis* as well—is spoken of *in two ways* [*zweifach*], for 'privation' [*Beraubung*] too is something like appearance" (WBP, 291–92/224–25).[23] We have already shown in what way *morphe* is *physis*, but what is new in these two lines is that privation (*steresis*) is something like appearance (*eidos*). Heidegger demonstrates what this means through Aristotle's example of warmth and cold (WBP, 293–94/226). When one encounters the appearance of, say, a metal cup filled with hot liquid, in lifting the cup to drink, one can say of the cup that it is warm or cold. In saying that it is warm, a certain attribution is made concerning the cup: it appears or unconceals itself as warm. The attribution accords with the appearance, and this is typically what we take to be a correspondence theory of truth. However, in attributing warmth to the cup, in saying that it is warm, one also makes a concurrent denial (*Verneinung*) of certain attributions to the cup: it is *not* cold; the appearance conceals its coldness. Accordingly, we should not hear in the German word *Beraubung* another form of the Latin *privatio*, but rather we should hear the word in its other sense of something being robbed. In the case of the example, in attributing warmth, it is not the case that there is a lack of coldness in the cup (or if it appeared cold, that is lacked warmth). Rather, in the attribution of warmth to the cup, another property is robbed from it; coldness does not presence, the warmth robs the cup of its potential coldness. Heidegger adds an important qualification to this conclusion that *steresis* is "absencing" in order to avoid the association of an absent property with *privatio*: in *steresis*, the absence is itself present as an absence (WBP, 294–95/226–27). But what does this mean? Examples of this have abounded since Heidegger, ranging from Sartre's example of searching for Pierre in a café at which Pierre has not yet arrived, to Derrida's spectral trace. As we will discuss at length in the fourth chapter, Merleau-Ponty makes this absence that is a presence the very foundation of his philosophy of expression, a philosophy that will attempt to unify nature, life, and history as interwoven "leaves [*feuillets*]" of being.[24] What is important is that the property is present, but present as *concealed*. *Steresis*, then, is "*something like*" *eidos*; it is not unconcealed as enframed in appearance, and yet it does manifest in the appearance as concealed. Consequently, when Heidegger claims that the "hidden essence of *physis*" is recovered in this thought of *steresis* (WBP, 295/227), we see what has been forgotten within our thought of nature: rather than taking natural properties to be based upon or inhere within a fully present substance, properties are always relational, that is, properties are unconcealed in an appearance only with respect to a meaningful world that conceals other properties in the process of unconcealing others. Returning to the example, perhaps when I lift the cup it is too hot for me to bear and I put it down immediately. The appearance in this case does not unconceal a drinking vessel, but rather a source of pain. The intensity of the heat, conceals or denies the cup's everyday meaning of "drinking vessel" for the moment, but by no means does the cup become any less a drinking vessel. This everyday meaning is still present, even as it is concealed relative to the heat. Another person with a greater tolerance for heat may lift the same cup without trouble, and in this case the relations of what is concealed and unconcealed are reversed: the cup as a possible source of pain is covered over by the everyday meaning of drinking vessel. We are meant to see in this interplay between concealment and presencing an addition to the traditional conception of *physis*: in any manifestation, in any enframing in presence, some possible meanings of a thing are made present at the expense of other possible meanings that are concealed through their being denied given the appearance that is unconcealed. What has been forgotten about the essence of *physis* is not only its *steresis* character but also its intrinsic relation to *aletheia*.

Through this forgetting of *physis*, Western thought becomes the progressive unfolding of this original responding to the call of being. This unfolding, however, is neither arbitrary nor independent of this original determination of *physis* because it is the conception of *physis* that has been described in this section that opens up the path for *physis* to be restricted to the sense of material nature, to be made the foundation for rational truth, and to become confused with and dominated by the modern distortion of another Greek concept, *techne*. Before beginning to discuss the effects of this history in our contemporary setting, we must, therefore, say a word or two about how *physis* and *techne* are related.

Heidegger's final determination of the dual essence of *physis* in Aristotle is actually given in response to a contrast with *techne*. He says the following: "*Physis*, on the other hand, is the presencing of the absencing of itself, which is on-the-way from itself and into itself. As such an absencing, *physis* remains a going-back-into-itself, but this going-back is only the going of an emergence [*Aufgehens*]" (WBP, 297/228). Understanding *physis* as a movement of emergence returns here with the now explicit connection of this movement of manifestation with a process of movement between various appearances. As emergence, however, we must be cautious not to understand *physis* as some kind of self-fabrication, something self-caused, as in the modern sense of the organism. The reason for this danger is that regarding *physis* in this way invites confusion with another kind of movement, *techne* (WBP, 253/195). For Aristotle, but unlike the pre-Socratics, both *physis* and *techne* are forms of *poiesis*, ways of being brought forth. The difference between these two kinds of movement is that *techne* has its origin *outside* itself rather than being its own origin—the first sense of *physis*. The movement of *techne*, then, will require harnessing the movement of *physis*, which is to say that all *techne* is a certain kind of knowing of *physis* that allows for *physis* to be put to use to a certain purpose. In other words, in technical knowing a certain appearance emerges precisely because *physis* is guided in advance by the *idea*, which will bring into presence a certain *eidos* that remains possible or latent within *physis*. The distinction here is twofold. First, *physis*-beings have the origin of their movement within themselves, while *techne*-beings have the origin of their movement in the *idea*. In this way *techne*, for Aristotle, is always going to be a derivative of *physis*. Since *techne* guides *physis* in the production of an appearance, there must be some kind of external imposition upon the movement of *physis* in order for the appearance to be produced such as it is. In the case of *techne*, *physis* would not have produced the appearance in the manner in which it was produced without the intervention of the technician. Take, for example, a chair made out of wood. There is no movement of the wood out of which the chair is made that would have brought about the genesis of the appearance of a chair. Yet, the chair is made of wood and is such due to an externally imposed movement that is still in accord with the movement of the wood. If the imposed movement were not in accord with the original movement of the wood, the chair could not be produced out of wood. Thus, there is a distinction between two kinds of beings: those that are brought about from out of the being itself (the movement of *physis*), and those that are brought about from out of a knowing of *physis* that harnesses it for specific purposes (the movement of *techne*) (WBP, 288/221).

What the forgetting of the *steresis* character of *physis* makes possible is the ascendancy of *techne* over *physis*. Once the relation of *physis* to *aletheia* and *steresis* is lost, it becomes possible to disjoin *physis* from *morphe* and rather associate it more closely with *hyle*. The reason for this is that *steresis*, the concealed or denied properties, are thought to originate in *hyle*, taken as matter that itself is the origin of the "potentialities" of a being. Here we hear the return of the more conventional associations of *morphe* with actuality and *hyle* with potentiality, resulting in the covering over of the relation of *physis* to *aletheia*. For in the case of the conventional translations of these two words, a being becomes a "composite substance" with form imposed upon a matter.[25] This conventional view lends itself to thinking of *morphe* and *hyle* as independent of each other, as if there could be a being that was a matterless form or a formless matter. Aristotle, of course, rejects these latter notions.[26]

Having arrived at this point, we must take stock of the road we have traveled in order to see more clearly why we have come this way at all. Perhaps the best way to do this is to look at Heidegger's understanding of *physis* as expressed in his interpretation of Heraclitus's fragment 123 as it appears in both *Introduction to Metaphysics* and as a foil to Aristotle's view in his reading of the *Physics*. In both instances he treats the fragment differently, but both treatments express the same understanding adapted to the context.[27] The fragment is popularly translated as "Nature loves to hide itself."[28] In *Introduction to Metaphysics*, Heidegger offers the translation of "Being (emerging manifestation) inclines in itself toward self-concealment" (*EM*, 123/121), which is also what he means in the later text when he asserts that the inceptual sense of *physis* is that of "being [as] the self-concealing unconcealment" (WBP, 299/230). In the inceptual sense, being is explicitly determined with respect to how beings unconceal themselves but at the same time remain concealed as well because, as we saw, every unconcealing also conceals. This understanding of being, of course, accords with Aristotle's definition of *physis* in the twofold sense. Why, then, is Aristotle's concept of *physis* but an echo of the thought brought forth in Heraclitus's fragment? Precisely because *physis* no longer remains the being of beings as a whole, but rather one particular kind of *poiesis* among others, for example, *techne*. The transition of sense in the meaning of being from *physis* to *poiesis* creates the possibility of distinguishing "natural" beings from "artifactual" beings, which further enables us to ask what these two kinds of beings have in common, that is, what is the being common to both. Here, Heidegger suggests, we begin to take *physis* not as a movement, but in its sense as constant presence, which he argues permits the creation of hierarchies within being.

The purpose of examining Heidegger's interpretation of *physis* in the Greek world is not to assess the accuracy of Heidegger's interpretation of Aristotle and Heraclitus. Rather, what is of primary concern to us here is the context it provides for understanding his views about technology and history. If the guiding question of Western metaphysics—the question that has been continually asked in Western thought, the responses to which constitute the history of that tradition—is, "What is the being of beings?" and "*physis*" is the initial answer to that question, understanding Heidegger on this point brings us to a new sense of historicity. Historical change becomes a matter of the *world*, in his proprietary sense of the relational totality in which we always already find ourselves, changing. *Thinking*, reflection upon the basic concepts that constitute the world, then becomes the motor of history because only thinking calls us to reconsider our answers to the guiding and grounding questions (more on the grounding question shortly). Showing how the Greeks understood *physis* to have a twofold essence involving both *ousia* and *steresis*, and that *physis* was conceived as something internal to a body rather than an external framework of laws, shows us that *a different way of collecting bodies is possible.* Another way to think about history, to put the matter in Latour's language, is the series of transitions in how propositions are collected and therefore which propositions are no longer recognized, which are excluded, and which are admitted. As we will see, Heidegger's views about technology that Latour resists so fervently are more concerns about how the manner in which a certain way of regarding beings that is proper to technology collects bodies than it is a rejection of technology as a whole. It is toward explaining this aspect of his thought that we must now turn our attention.

III. THE RISE OF TECHNOLOGY AND THE DEVASTATION OF THE EARTH

The critique of the technological worldview and Latour's criticism of it are best understood in terms of the history of being that was introduced in the last section. Having completed our foray into Greek philosophy, we can now turn our attention toward the legacy of that thinking: how *physis*, as the word for both being and the being of beings, became "nature" in the modern world. Because these conceptual relations were developed in response to the guiding question, which was put to the thinker by the world, they are still with us today in various ways. It is this conceptual remainder, the ways in which we remain within the domain of the initial inception of philosophy without questioning whether we might think of that beginning differently, that we must bring to the fore in order to assess the extent to which Heidegger's philosophy and Latour's are compatible and the extent to which Latour's criticism of Heidegger's views on technology capture the central concern of those views. Without entering into his account of this history of being in too much detail, we can begin this process by turning our attention toward the evolution of the grounding question of Western metaphysics—namely, "What is being as such?"—as a distinctive question. As I will argue, Heidegger's real concern about technology is that it occludes that the guiding question is itself even a question and therefore makes possible the transition to an anthropocentric worldview, wherein nature becomes a resource for human consumption. As mentioned in the introduction to this chapter, however, this effacement of the guiding question and consequent elimination of the grounding question are what brings our attention to the fact that these are questions to which we provide answers. As Heidegger puts it rather pithily, "The closer we come to the danger, the more brightly do the ways into the saving power begin to shine and the more questioning we become" (FT, 35/34). If it were possible to relate to technology differently and allow the grounding question, then many of the issues with technology could be alleviated.

A response to the grounding question establishes a foundation from which responses to the guiding question are offered. Heidegger calls this progression of interpretations of the being of being on the basis of an unquestioned determination of being as such a "historical sending." Heidegger gives one of his clearest explanations of the historical sending resulting from Aristotle's conception of *physis* in "The Question Concerning Technology," which is incidentally the text that Latour most frequently criticizes. In this essay, Heidegger argues that every kind of unconcealment comes about on the path opened by a definite disclosure of beings (FT, 25/25). With respect to our previous discussion, we can understand this assertion in terms of how the determination of being as *physis* unconceals beings as a whole in a specific manner, as constantly present. Recall that beings are always disclosed within the context of a world, meaning that the contours of a world can be approached by examining the manner in which participants in that world disclose beings to themselves and therefore interpret their environs. For example, if, within the Greek world, the being of beings is taken to be *physis*, then certain characteristics of beings will be unconcealed in terms of specific meanings that would perhaps be concealed if the being of beings were interpreted otherwise. More specifically, something like Aristotle's conception of "natural slaves" can be contextualized (though I am not saying justified) by an understanding of *physis* and violent movement: slavery is not unjust because it is not a position imposed and sustained by force; there are people for whom their *physis* is to occupy what was considered to be a necessary role in Athenian society. In the process, what are today obvious ways of regarding the so-called natural slaves are concealed to the Greeks due to the world in which they dwell.[29] The determination within a given world of what the being of beings is, then, opens up a *way* of disclosing beings, which is what Heidegger means when he speaks of being "upon the way [*auf einem Weg*]." With the conception of the being of beings, a people is sent (*schicken*) down a path of disclosing that opens up a destiny (*Geschick*) for interpreting beings (25/24). "Destiny" should not be heard in the sense of a predestination or unavoidable fate, however. Rather, we should pay heed to the resonance "destiny" takes on within a world. In any given world, one is *destined* to have beings disclosed or unconcealed to one in a specific way. If one does not question the meaning of being itself or metaphysically question how beings are interpreted in their being, this pregiven mode of disclosure will remain in place. Nevertheless, this destiny can be displaced through questioning how the path upon which one is sent came to be in the first place, that is, by questioning the meaning of being. For this reason, we examined the account Heidegger offered of the original Greek historical sending in the thought of *physis*. We saw there that *polemos* opens up authentic history, and in this context the idea is developed further such that "destiny" determines "the essence of all history" (25/24). In the context of the earlier discussion, we can see how the conflation of being with the being of beings holds open a certain path along which beings were destined to be interpreted within a certain range. History, then, is the unfolding of the Greek path of thought and the transformation of this path into something that the Greeks would most likely never recognize.

Since we no longer live within the Greek world, though, it would serve us well to ask how being is thought of today. Just as the path of *physis* concerned a specific manner of disclosing beings, so, too, will the path upon which we find ourselves. The question, then, is How are beings disclosed within the contemporary world? Heidegger's answer to this question is that beings are disclosed to us as *Bestand*, which we have already mentioned is usually translated as "standing-reserve." While this is a fine translation of the word, it is also interesting to note the temporal sense that does not clearly appear in the English: *Bestand* also has the sense of "persistence," a being's enduring in time. Heidegger is taking up this sense of the word, *along with* the sense of a reserve that is always at hand, in order to call to our attention how contemporary notions of time and temporality then lead to specific disclosures of beings. This line of thought is reminiscent of Latour's views in *We Have Never Been Modern* with respect to how modernity requires a specific conception of linearized time in order to function in the way it does (*NBM*, 67–76). They both argue for an alternative to the modern conception of history in which the past is borne in the present and the reversibility of time is rejected. The passive notion of *Bestand* contains many of the notions about objects that Latour himself finds distinctive of modernity.

What leads to the revelation of beings in terms of *Bestand* are the challenges (*Herausfordern*) that technology makes possible. "Challenges" in this case refers not to difficulties or problems, but rather to specific kinds of demands that are made upon both the contemporary human being and the surrounding world of the human being. Heidegger's example of a farmer is illuminating of this point (FT, 15–16/14–15). The traditional work of the farmer does not necessarily challenge the land: seeds are sown, fields are tilled, and the processes of the land issue forth crops. Certainly the work of the farmer orders (*bestellen*) the land, but the farmer does not ask the land to do anything outside its cycles of growth and rest. Within a disclosing of beings that is characterized by challenging, however, the farmer's ordering of the land takes on a new aspect. The land is no longer that *with* which the farmer works, but that *upon* which the farmer works, applying the latest machinery, GMO crops, and chemical or organic pesticides in order to *maximize* the yield, in order not to *waste* any of the land's potential. The demand to maximize and not to waste any latent potential in the land is a *challenge* to the earth; the modern farmer pushes the earth to its limits and possibly beyond. The issue is not necessarily the kind of practices that take place on the land; in both cases, farming remains the planting, tending, and harvesting of crops. What has changed is the way in which the farmer relates to the earth and thereby is able to demand more and more of it. In this example, the relation between the farmer and the earth is what subsequently allows the earth to be dealt with in an exploitative way. This shift in relation is even more apparent if we imagine that, beneath this tract of farmland, there is discovered a valuable resource such as oil or a rare earth metal. The earth in the field in this case will even cease to be viewed as earth at all, becoming disclosed as a depository for that resource, an obstacle in the way of extracting it. The soil recedes into the background and is no longer disclosed as a complex, organic phenomenon. What happens in both cases is that there is a change in worlds: beings are no longer seen as having the *arche* of their movement within themselves (*physis*), but rather they are seen as having the meaning of their being outside themselves in their instrumentality. Since instrumentality is defined from the human perspective, human beings then assume the role of orderers of what is disclosed in the environment. This specific form of gathering-disclosing Heidegger calls "*Ge-stell*," enframing (FT, 20/19). We can here note, however, another similarity between Latour's view and Heidegger's: what is at issue is a specific way of gathering beings together and how those beings are interpreted.

Heidegger calls the relationship to beings that institutes the technological world "machination [*Machenschaft*]," the operative root of which is *machen*, which not only means "to make" or "to do," but also maintains a relation to *Macht*, meaning "power." Machination, then, comes about on the path of disclosing opened by *techne*, itself a form of disclosing that was said to be a kind of knowledge that guides *physis*. Rather than knowledge of how a being emerges guiding the process of emergence toward the manifestation of a specific end (e.g., the traditional farmer), within machination the being is put to work in a way that expands the limits of its power (e.g., agribusiness that incorporates genetic engineering) (*BP*, 109/76). The expansion of limits is what we called "challenging" above. Even though Heidegger does understand machination to be an "'evil' type of human activity and plotting for such an activity" (126/88), we must be cautious to understand it in the proper way so as to understand where the evil lies. Machination is *not* something bad because it distances those within its mode of disclosing from the correct understanding of being. Just as *physis* is one way of understanding the being of a being, machination is nothing more than another way of understanding beings. As a moment along the same path upon which *physis* placed Western thinking, machination comes to prevail through the same intuition that determines being as constant presence (126/88). The transformation from *physis* to machination takes place, as alluded to above, when *techne* and *physis* are both thought under the heading of *poiesis*, a bringing forth into presence, a making (*machen*). Understood in terms of *poiesis*, *physis* no longer retains its sense of a self-emergent manifestation. Rather, *physis* becomes understood as a being making itself, a change that requires a maker. This maker is fabricated within the concurrent change in the understanding of *ousia* from enframing in presence to *substantia* or substance: *physis* becomes *natura*, a domain of laws that cause a being's substance to behave in specific ways. The dominance of *poiesis* in the Christian world takes the form of an interpretation of beings such that they are said to be caused or created as opposed to self-emerging (171/120). Once beings are seen as caused, the next step along the path is to see them as ready at hand, where their instrumentality comes to the fore and beings are disclosed as *Bestand*. As the latest development in the movement initiated by the ancient Greeks, machination arises out of the previous strife between manifestings that holds open the world. What all three transitions in disclosing—from *physis* as internal principle, to *substantia* as what is formed, to *nature* as external system—have in common is their agreement that what makes a being a being is the constancy of its presence. Each of the modes of interpreting the being of beings simply takes this constancy of presence to mean something distinct. The reason behind this commonality is that the grounding question had not been explicitly addressed; Heidegger's analysis is meant to call our philosophical attention to this and have us ask whether constant presence is the best way to understand being.

Machination, as a mode of sending, is dangerous, but it is important to emphasize that this danger does not come from the content of the practices within it per se. Rather, what makes machination dangerous is in its sending forth of humanity toward a destiny. Heidegger explains this succinctly in "The Question Concerning Technology" when he describes "the highest danger" posed by enframing (FT, 27–28/26–27). The danger of any sending lies in the interpretation of beings becoming confused with truth. When a being is confronted within a world, the being is always encountered *as* something. The "as," however, is determinable only in terms of the preunderstanding of being, that is, the interpretation of the being of beings, that prevails within that world. Consequently, certain statements concerning the being can be correct *given that interpretation of beings.* Given another interpretation of beings, within a different world, the same statement made of the same being could be incorrect. Correctness and truth, insofar as truth is essentially related to unconcealment and concealment, are not to be confused. In "On the Essence of Truth," Heidegger explicitly ties the truth of a being to a "mystery:" truth is both what is unconcealed in any given relation *along with* all those meanings that are concealed by that unconcealment. Since it is not possible to disclose all the possible relations to any given being in a single manifestation of the being or even exhaustively over a period of time, the essence of the being or its truth remains mysterious. In pursuing the complete disclosure of a being from a given interpretation of beings, the mystery is forgotten and correctness comes to stand in the place of truth. Since unconcealedness occurs only in and through relations, the capacity to enter into myriad relations with a given being, freedom, is the essence of truth. If humanity is free through its capacity to wander or err along various paths upon which it is sent, then, in the forgetting of the mystery and the attendant displacement of truth in favor of correctness, humanity loses its freedom. The sending is dangerous, then, insofar as it leads toward the loss of humanity's openness to thinking about the meaning of being due to a loss of the freedom that would enable such thinking.

The historical sending of machination becomes "evil," however, through both instilling the will to mastery and leading us to forget that there are indeed other ways of interpreting the being of beings. As an interpretation of the being of beings, machination is *not* a deviation, a mistake to be corrected, or, more importantly, a falling away from the original interpretation of *physis* that, were humanity to have maintained this interpretation, would have somehow prevented the ascendance of machination. Heidegger enumerates three "laws of machination" to help make sense of why machination is a disclosing of a different kind than *physis*, even though the two are intimately related (*BP*, 127/89). The three laws are as follows:

1. "The more authoritatively [*maßgebender*] it unfolds . . . the more stubbornly and more machinatingly it conceals itself as such . . . as basic forms of actuality and thus of beingness."
2. "The more decisively machination conceals itself in this way, the more it rushes toward the predominance of that which seems to be totally against its essence and nevertheless is its essence: *lived-experience*."
3. "The more unconditionally lived-experience becomes the measure [*Maßgabe*] for correctness and truth (and thereby 'actuality' and constancy), there becomes no chance of fulfilling, from here, a knowledge of machination as such."

We already have the groundwork for understanding these three laws and their import through the structure of sending and destining described above. Here we see exactly *how* machination sends humanity on the way. In the first law, the assertion is that machination discloses beings in a certain way, but, in this mode of disclosing, it conceals the way in which it gives the measure (*maßgeben*) to beings. In other words, it becomes *obvious* that beings are present at hand and defined by instrumentality, an obviousness that becomes increasingly unquestionable. If machination is a disclosing, the concealment of its measure-giving (its authority) amounts to a concealment of the relational character of the beingness of beings, a concealment of how beings come to be interpreted through a relation with them. This establishing of a single track for disclosing beings is what Heidegger means when he speaks of dominating them: the mystery of being is ignored or forgotten, and the one disclosure therefore limits a being's essence to a single one of its possible manifestations. Accordingly, a being's "usefulness" comes to be seen as a property of the being itself, not as a consequence of a relation between two beings; and a being's instrumentality becomes the scale upon which its value is judged. Here is where the will to mastery emerges, and this tendency is already witnessed in figures like Descartes: the purpose of knowing is to put beings to use for us in ever more challenging ways. The second law follows from the concealment of machination as a mode of disclosure: if the relational character of unconcealment is itself concealed, then machination begins to take hold not only among the beings with which humans relate themselves but also among human beings themselves. In another context, Heidegger defines lived-experience as "the basic kind of machinational representing and the holding of oneself therein" and "the publicity that is accessible to everyone of the mysterious, i.e., the stimulating, provocative, stunning, and enchanting, which the machinational makes necessary" (*BP*, 109/77). Though it would take us too far afield to assess the validity of Heidegger's charges against the Husserlian phenomenology, we can see Heidegger distancing himself from the subjectivistic interpretation of phenomenology Latour criticized above. What is at stake is whether lived-experience serves to oppose or to reinforce the machinistic way of disclosing beings. Heidegger's claim is that lived-experience becomes a valid standard for truth only when beings are already being disclosed through machination. The conception of lived-experience entails an interpretation of objects as passive and constantly present to our perception, the exact conception of the object Latour criticizes; objects are always intentional objects there for interaction with human consciousness and use. Hence, in opposing machination and rejecting lived-experience as a foundation for that opposition, Heidegger can reasonably be said to be moving toward a more object-oriented philosophy, though we will see that he has not fully achieved his goal in the next chapter. The tendency to reduce the object to an intentional object leads to the third law of machination. According to this law, lived-experience, how a human being is disclosed in machination, comes to be the measure for truth, thereby cutting off the originary mystery that is truth and replacing it with the correctness-correspondence theory of truth that arises out of an improper understanding of the relation between *physis* and *aletheia*.

These three laws are reiterated in a briefer form within "The Question Concerning Technology" in a way that makes their relevance to our current analysis more salient. The danger posed by the sending of enframing confronts humanity on two distinct fronts. First, there is the direct danger to humanity itself: as beings within the world, human beings must also view the being of their being as consisting in *Bestand*. In other words, human beings become another resource at hand for the use of human beings. Human beings set to ordering themselves and their societies in the same way that "objects" are ordered and organized so as to be challenged into *producing*. At the same time that humanity is reduced to a mere resource, the second threat arising from machination paradoxically sets humanity above all the other beings (FT, 28/27). This ascendancy is due to humanity's place as the *orderer* of beings, the one for whom all beings are present: humanity becomes "lord of the earth." The relationship to the earth as *Bestand* opens up the possibility for exploitative practices, which is also not to say that machination offers the only interpretation of the being of beings that would lead to such exploitation of the earth.

But how is nature—and here Heidegger himself is using the word *Natur*—thought of when it is unconcealed as *Bestand*? What nature places at hand, what makes nature eminently *useful*, is energy (FT, 22/21). Discussions today concerning the availability of energy are commonplace, and the question of the accessibility of energy is frequently put in such a way that one must choose between either natural beauty or the survival of human beings. The problem is that, regardless of which choice is made, nature may still be taken as a use-object; each option construes nature as a source of energy in a different form. For example, we can look at two different national parks, Arctic National Wildlife Refuge (ANWR) and Yellowstone, both of which have very different relationships with human beings. It could not be feasibly proposed politically that Yellowstone be used as, say, the site for a geothermal energy plant because too many people derive enjoyment from the natural beauty of the place.[30] There is an already existing use for Yellowstone that would conflict with its new use as a power plant. The enjoyment-use, however, is just another way of exploiting the energy of a place for human ends. In ANWR, however, the situation differs insofar as its natural beauty is so far removed from the vast majority of people that this enjoyment-use never factors into the reckoning. ANWR can supposedly be mined because no one is using it, and (it is claimed, at least) no animals who do use it will have their use of it disrupted. In other words, we must drill there or else the land's potential is wasted.[31] Even in many environmentally minded tracts we still read about how nature is a circuit of forces, a network of energy in need of preservation. The disagreement between the advocates of this view and those who would put nature to use is not the conception of nature itself, but rather that for which nature is intended: human use or the ecological flow of forces. The confluence of these two camps' respective understandings of nature can be seen in their mutual reference to the science of ecology in supporting their views: both accept the original definition of nature as a network of measurable and manageable forces, as a system. The argument from Heidegger is such that in this context even the approach of the ecologically minded individual remains within the disclosing of nature associated with machination through modern technology. Here is another point of connection with Latour: they both believe environmentalists' resistance to the exploitative practices of modern technological capitalism is doomed to reinforce the disclosing that allows for the exploitation as long as it operates within the enframing disclosing of beings. Though Latour does not discuss the modernist treatment of nonhumans in that language, we saw in the first chapter that he rejects exactly this way of regarding nonhumans as passive and exploitable.

Before ultimately adjudicating their dispute, however, it remains for us to explain the relationship between technology and science as it pertains to nature. Within machination, nature becomes an orderable and calculable system (FT, 24/23), and it is this notion that underlies, as for Latour, the metaphysical naturalism of Science. What is relevant to us is the manner in which the scientific sense of the system of nature then makes possible the mode of disclosure called enframing. Heidegger establishes this connection by transplanting a traditional phenomenological interpretation of the sciences and their reliance upon "regional ontologies" into his own ontology. Heidegger is frequently thought of as antiscientific or romantic, yet he does not dismiss science. He does, however, severely curtail the realm in which the epistemological conception of Science Latour also criticizes can be called upon to arbitrate what is and is not real. His views with respect to science, which are extremely nuanced and therefore cannot be rendered here in all their detail, are relevant to us to the extent that they contextualize the genesis of the idea of nature as a system.[32]

In *Being and Time*, science is described as one of the essential projects of Dasein, with Heidegger adding that phenomenological ontology is the search for the a priori conditions of possibility for the sciences (*SZ*, 11/9). Science is a search for knowledge concerning beings and as such begins from a situation of requiring an interpretation of beings.[33] The interpretation of beings provides a *Grundsätz*, a fundamental or grounding premise, which allows a science to choose out the beings that do and do not belong to its ordained region. In having the region so defined, science then proceeds to describe those beings in terms of the already decided upon grounding proposition through the application of a method.[34] What distinguishes a science is its method, and in seeking a perfect, totalizing method, science always approaches objects from "a determined direction of explainability [*eine bestimmte Richtung der Erklärbarkeit*], which already basically [*grundsätzlich*] guarantees a 'result' " (*BP*, 146–47/102). The word to take note of here is *Richtung*, which comes from the same root as *richtig* (correct), and *Richtigkeit* (correctness or truth). The implication is that the sciences operate within the predetermined region of beings such that they generate knowledge in the form of experimental results that are correct in light of the preinterpretation of being that determined the region in the first place. To reiterate, truth must still be reserved for the unconcealment-concealment structure, and thus a science never has access to the true because it always deals with beings as interpreted and never with what remains hidden or concealed in that interpretation. What Latour calls the sciences, insofar as they operate independently of the epistemological conviction of a preexisting and ahistorical natural order, would not fall prey to this problem. Though Heidegger later eschews this language within which fundamental ontology is said to be seeking the conditions of possibility for anything (*BP*, 239/169), he does retain the view that the sciences operate in terms of regional ontologies. In short, the sciences *clarify* and *explain* a predetermined object for their inquiry in terms of a prior interpretation of the being of beings (e.g., *BP*, 145–46/101). If the idea of nature as an orderable and calculable system is the fundamental or grounding idea of ecology as a science, and if that idea of nature is partially responsible for the devastation of nature, calling upon ecological science as the ultimate authority in how humanity ought to relate to nature is not necessarily going to remove us from our currently problematic set of relationships and behaviors. I do not mean to imply that there will be *no* use for ecology, only that, like technology, we will have to arrive at a more nuanced understanding of how to utilize ecological findings in our deliberations rather than appealing to those findings as the sole and unequivocal authority on matters concerning the environment.

Since the question we are dealing with is not one of science per se but rather concerns how the thought of nature as a system within Science is related to machination and its consequences for the earth, we still need to establish that machination sanctions the devastation of nature.[35] Machination, through its constant retreating from view, through its concealing itself in and fostering the development of lived-experience, and through the impossibility of coming to terms with machination from the perspective of lived-experience, increasingly stifles the strife between earth and world (*BP*, 277/195). This inhibition occurs, however, through the representation of nature as dually transcendent and immanent within Science. If nature is taken to be an object or a domain of objects for scientific inquiry, then it must also be placed within the appropriate region of scientific inquiry, in this case mathematical ecology. Within this representation of nature, the fundamental premise is taken as given or preunderstood, thereby ceasing the originary strife or *polemos*, which in turn allows being to appear as a being. Without the strife, the world is no longer held open, which is to say that there is no more room for the transformation of meanings within the ensuing system of relations. In terms of nature within the world of machination, the closing of nature within the representation of *Bestand* leaves nature as a realm of mechanical causes or teleological goals to be used as energy. Once everything "natural" has been reduced to *Bestand*, including human beings, the earth can be challenged further and further until humanity regards itself as responsible for ordering and organizing the whole of nature.

IV. CONCLUSION

At this point, we should briefly turn our attention back to the relationship between Heidegger and Latour in order to bring this critical phase of the argument to a close. Though I have been indicating throughout those points on which the two philosophers concur despite overt differences in their presentation and philosophical aims, it may be helpful to conclude this chapter by engaging with their disagreement in greater detail. Latour himself credits Heidegger with seeing the point at which the modern constitution holds together, namely, the denial of mediation in favor of dualistic purification, but immediately faults Heidegger for his view that "being cannot reside in ordinary beings" (*NBM*, 65). This reproach can be heard in several ways, and Latour does not clarify what he means exactly by it. There is one sense in which it is clearly false: things will continue to exist independently of human beings. Heidegger's view is not a subjective idealism. Though there is another sense in which it is definitely true: Heidegger believes that human beings are distinct from all other living beings in having a relationship with being itself and not being immediately absorbed in one's relationships; inanimate beings are even worse off insofar as they do not even have relationships in any meaningful way.[36] This view is explicit in Heidegger, but it is difficult to see how Latour can be said to reject it himself. Consider the following passage in which he is discussing the role of humanity in mediation:

> The expression "anthropomorphic" considerably underestimates our humanity. We should be talking about morphism. Morphism is the place where technomorphisms, zoomorphisms, phusimorphisms, ideomorphisms, theomorphisms, sociomorphisms, psychomorphisms, all come together. Their alliances and their exchanges, taken

> together, are what define the *anthropos*. A weaver of morphisms—isn't that enough of a definition? The closer the *anthropos* comes to this distribution, the more human it is. The farther away it moves, the more it takes on multiple forms in which its humanity quickly becomes indiscernible, even if its figures are those of the person, the individual or the self. By seeking to isolate its form from those it churns together, one does not defend humanism, one loses it. (*NBM*, 137)

While this passage is clearly meant to target Heidegger given the immediately following question of how machines challenge our basic humanity, the passage itself is a complete affirmation of Heidegger's central model. Humanity is defined through the special kinds of relations it takes up with the beings it encounters. The figure of humanity simply *is* the sum of the forms it uses to interact with beings, whether that form be naturalizing, socializing, psychologizing, et cetera: "Human nature is the set of its delegates and its representative, its figures and its messengers" (*NBM*, 138). It remains the case, however, that nonhumans will also be able to be defined in this way, which is something Heidegger would most certainly reject. Aside from this last point, however, his objection to Heidegger is simply that Heidegger believes that machination puts us on a path of disclosure that removes our ability to interpret beings diversely. In Latour's view, technology is not reductionistic in the same way as Heidegger conceives of it. Technology neither compels us to reduce nonhumans to their use value for humans since, as constituent parts of actants, nonhumans never lose their powers to actively contribute to the world. Of course, the problem with the modernists' constitution is that it ignores this contribution, which is similar, though not identical, to how Heidegger views the technological worldview as leading us to ignore the mystery. So in both cases, there is a failure of hermeneutical perception: what is at stake in challenging mastery is *how we view nonhumans*.

Another similarity lies in their respective views on mastery. As we saw in the last chapter, Latour believes that any attempt on the part of humanity to master beings will recoil upon us because it simply is not possible to have the kind of total control that certain ideals of mastery seek to attain; mastery, if possible at all, is a localized and very small-scale phenomenon. A problem with the project of mastery, in addition to embodying the aforementioned failure of hermeneutical perception, is that it is itself impractical: there is more potential for understanding agency in viewing a multiplicity of humans and nonhumans collaborating as an actant to bring about certain effects than there is in trying to isolate a single being that can take primary responsibility for a given effect. This view is what leads Søren Riis to say that Latour's philosophy is itself an embodiment of enframement, but it is important for us to estimate to what extent this charge is valid.[37] It is incontrovertible, both Riis and Jeff Kochan point out, that for Latour all beings are members of larger networks of relations.[38] Both of them, however, believe that the mere fact that Latour thinks in terms of networks is sufficient to make his views subject to Heidegger's criticism of enframing. This criticism, however, is just as unfair as some of Latour's one-liners directed against Heidegger. Enframing entails foreclosing alternative possibilities for interpretation of beings by reducing them to their existence as potential energy. If there is anything that is clear from our previous reading of Latour, however, it is that one of his primary concerns is *to keep open* the possibilities for collecting nonhumans, to ensure that there are always more than two ways of doing so, and to allow nonhumans to speak in their own ways within a democracy of objects. So even if Latour does describe his philosophy of science and technology in some of the terms Heidegger critiques, his system still does not fall prey to the tendency to mastery Heidegger rejects, since he also liberates science and technology from the modernist conception of nature. The motivation behind the rejection of nature would also lead to a rejection of enframing.

So who is right in the dispute between Heidegger, Latour, and their followers? I hope to have shown that neither is completely correct and that each has something to learn from the other. Heidegger can learn from Latour that nonhumans can be much more active than they are frequently given philosophical credit for. This intuition will be explored most closely in the fourth chapter, as it has rather dramatic consequences for phenomenology in general. Latour, on the other hand, can learn from Heidegger that, as Kochan suggests, mediation itself has a history and is not a transhistorical structure.[39] Since the being of beings has been interpreted diversely throughout Western—not to mention non-Western!—history, what constitutes mediation and how mediation functions will be historically relative. The historicity of mediation does not seriously call into question Latour's views, however; if anything, it strengthens them, as it merely introduces another set of relations that his relationism needs to take account of.

The last point I would mention here is something about which both Heidegger and Latour might be equally uncomfortable: rather than abandoning nature altogether, it may be more productive philosophically to retain *a sense* of nature in a transitional way. Since nature has been such a foundational concept in Western thought since its inception in Greece, we cannot simply reject it because we now recognize that the way in which we have conceived of it has been misguided. Rather, and this is what I will begin to discuss in the next chapter, it would be better to "twist free" of the concept by reconceiving it in a way that makes it less central and more escapable than previously. If nature can be a point of connection between individuals and the surrounding environment and at the same time retain the nontotalizing structure described in the first chapter, then it may yet serve a role in the nonmodern world.

CHAPTER 3

THE OPENING OF THE EARTH

> Thus we ask now: even if the old rootedness is being lost in this age, may not a new ground and foundation be granted again to man, a foundation and ground out of which man's nature and all his works can flourish in a new way even in the atomic age? . . .
>
> If *Gelassenheit* toward things and openness to the mystery awaken within us, then we should arrive at a path that will lead to a new ground and foundation. In that ground the creativity which produces lasting works could strike new roots.
>
> —Martin Heidegger, "Memorial Address"

If we are to turn aside the view of nature as a standing-reserve of energy, as the vast majority of environmentalist literature today professes a desire to do, the suggestion received from Heidegger via his analysis of technology and science is that the originary strife that holds the world open must be renewed and we must *hold open the mystery*. For Heidegger, the originary strife is between earth and world, which, unlike the binary pairs of concepts prevalent within modernity, do not correspond to the categories of nature and culture and in important ways cannot be separated from each other. Heidegger's strife, in order truly to open up a questioning of being, avoids dualistic construction by not allowing the originary terms to be defined in an oppositional manner. How, then, does Heidegger define the terms in originary strife? Answering this question will also allow us to see what becomes of nature in Heidegger's thought. With all these critical elements introduced, then, the challenge that remains for this chapter is to determine what role nature still may play within ontological thinking.

In beginning this chapter, it may prove helpful to recapitulate exactly where we stand with respect to Heidegger's thought from the previous one. In the first section we dealt primarily with his attempted recovery of the first beginning of philosophy through an encounter with the Greek notion of *physis*. There, we saw *physis* determined in terms of *Aufgehen*, the emergence or rising up into manifestation. Invoking Heraclitus's fragment 53 and Aristotle's *Physics* B, I, Heidegger shows how *physis* is involved in an original *polemos* that holds open a world such that beings are identified *as* beings. Through this *polemos*, beings are gathered (*logos*) and thereby unconcealed (*aletheia*) in a determinate way that conceals or denies (*steresis*) other determinations.

This recovery of the first beginning, however, was done for the sake of clarifying the manner in which humanity was thereby sent forth upon a determinate path by the Greek thought of the being of beings. Thus, in the second section, we discerned our current position along this path and how it was that we arrived at that point departing from the inceptual thought of *physis*. Here we discussed the manner in which to be sent upon a path is dangerous due to its setting forth a destiny or historical sending for the interpretation of beings. Destining closes up the world by covering over the originary strife that holds the world open, thus fixing the meaning of a being to a specific quiddity. The particular destining that holds sway today we identified as machination, which determines the being of beings as standing-reserve. Rather than thinking of nature in terms of emerging and manifesting, nature is transformed into a determinate being at hand, like all other beings, and thereby into a use-object. This conception of nature as a use-object is what we now seek to confront and in so doing attempt to show how one might open up the possibility for a new relationship to nature.

Before that attempt, however, a moment must be taken to explain how we ought not to understand Heidegger's directives up to this point. We should not think of Heidegger's views with respect to the history of being as deterministic. On the contrary, he intends us to understand that machination is itself a historical possibility for the interpretation of beings like any other, and, as a possibility, one that could have been otherwise. For this reason Latour is incorrect to say that technology is, for Heidegger, "unique, insuperable, omnipresent, superior, a monster born in our midst which has already devoured its unwitting midwives" (*PH*, 176). On the interpretation of Heidegger's thought offered here, the challenge of his work is *not* to resuscitate the Greek relations to being so that we can today oppose the forces of machination with this earlier, purer view out of which it arose.[1] The view that Heidegger's philosophy is somehow "romantic" insofar as he believed that a nontechnological world would somehow be possible today is a mistaken one as well. Though his examples are frequently pastoral, there is nothing in the philosophy that would indicate a requirement to return to archaic ways of living as Latour frequently implies.[2] As Heidegger puts it, he desires "to prepare a free relation to [technology]," within which "our Dasein is open to the essence of technology" (FT, 7/3). As Kochan explains this point, "the attainment of a free relation to technology is possible only once the essence of technology becomes an object of critical scrutiny."[3] So it is not the case that technology qua technology is the problem. Heidegger's concern is with technology qua totalizing, modernistic worldview. Once the *worldview* of technology is broken through, we might be able to analyze technological innovations in terms of the particular relationships with other beings they inspire. This project seems to me profoundly in line with Latour's aims. Also mistaken are those views that assert that his philosophy can offer no resistance to the forces of machination because every possible interpretation of beings sends one down a path and this sending itself contains the danger of mastery over beings.[4] As we saw, Heidegger himself considered machination to be an evil and so, presumably, would attempt to develop a means by which to avoid it or, if unavoidable, at least to mitigate or confront this evil. In order to circumvent these misunderstandings of Heidegger's thought, two points are in need of clarification: the structure of decision and the relation between the first inception of philosophy and Heidegger's proposed "other inception." We will deal with the latter first.

I. DECIDING AGAINST THE SYSTEM OF NATURE

The entirety of the Heideggerian project is aimed at preparing for what he calls an "other inception." In the same way that the Greeks thoughtfully responded to that which was question-worthy in their times (the being of beings), so must we come to thoughtfully respond to what has become question-worthy in our own time (being as such). This distinction is, again, what Heidegger terms "ontological difference" and is prefigured in the philosophies of Leibniz and Schelling as well.[5] Of course, the task of responding to the grounding question is much more complicated today given the ways in which machination occludes the relation with being as such. This difficulty is the motivation behind the recovery of the first, Greek inception of philosophy: to uncover the manner in which machination came about as a result of a conflation of being with the being of beings enables us to ask the grounding question explicitly and respond to it differently. As Heidegger alternatively describes the first inception, it asked the question of what is true with respect to beings without ever questioning the essence of truth as such (*BP*, 179/125). In other words, truth remains *ungrounded*, or at least inadequately so, within the first inception in such a way that truth, as we have already seen, becomes correctness. When the guiding question of metaphysics is asked within the first inception, however, this is not a mistake and its history is not a fall: the positions taken toward the being of beings (and thus to being itself) in response to the guiding question all remain positive examples of how one may interpret the being of beings. Accordingly, we said the history of Western philosophy is the history of responding to the guiding question in various ways such that philosophers' interpretations of the being of beings have provided assorted representations of what nature, humanity, and society are like. They all inquire into the manner in which being holds sway or prevails (*walten*), without taking the additional step to ask how or why it prevails in that way. By looking at the ways in which philosophers have responded to the guiding question, the hope is that the dialogue with those thinkers will provide a vista toward how to *begin* to ask the grounding question and thereby open up the other inception.

In asking the grounding question, a new philosophical project is opened up, within which the goal is no longer to provide a novel interpretation of the being of beings, but rather to seek being (*BP*, 17/13). To reiterate, the goal is not any particular interpretation of beings, naming of being itself, or instituting any *one* relation to being: the questioning itself is the goal of the search because, in asking the question, the sending is never begun; the world is held *open*.[6] One cannot simply ask the question, however; one must first enter into a position from which it is possible to see why the grounding question is not merely one question among others, but *the* philosophical question. This position is attained through a *confrontation* (*Auseinandersetzung*) with the first inception, and with this idea we see that strife is possible, not only between manifestations, but also between worlds. Within a world, a given disclosure of beings allows for differences in status and rank. To say that there could be a confrontation between worlds is not simply to assert a conflict of values or civilizations, but rather is a confrontation between fundamentally distinct ways of articulating beings and their being.[7] In terms of nature, the importance of this confrontation that questions status and rank should be self-evident: if the idea of "nature" as an object predisposes us to understand it as standing-reserve, as something constantly present and merely instrumentally useful as energy, then displacing this conception of nature so as to give it a different status and rank is of primary importance in any attempt to rethink the relation of humanity to nature.

So the question becomes how Heidegger enacts this confrontation that will prepare for the other inception of thinking. The first thing to note concerning the confrontation is that it is not an opposition to the first inception. If everything that unfolds within metaphysics is an essential possibility of being, then it makes little sense to say that one is seeking the essencing (*Wesung*) of being and then set about denying that which metaphysics reveals as part of that essencing.[8] More concretely, if it is a specific interpretation of beings that one wishes to reject (e.g., one that takes nature as a use-object), merely offering another interpretation of beings (e.g., nature is a harmonious system independent of human use) is not sufficient to displace those attitudes of dominance and anthropocentrism that were so problematic in the first place. The oppositional fixation of beings in a counterinterpretation would merely be a countersending with its own "counterinstitutions" (*Gegeneinrichtungen*) that give it a "counterdirection" (*Gegenrichtung*) (*BP*, 186–87/131). Another way of understanding this leaving behind of opposition is through the two terms *Gegeneinrichtung* and *Gegenrichtung*: in both cases opposition or countermovement (*Gegenbewegung*: note the *Weg* and hear "an opposed being on the way") establishes another correctness that corresponds to the interpretation of beings to which it is opposed, and this correctness is guaranteed through institutions. In other words, the other inception cannot be opposed to the first because such opposition would offer nothing more than another way of mastering beings.[9] If the goal of the confrontation is a seeking questioning and not an interpretation, however, the attitude of mastery dissipates in the lack of a *definitive* or *predetermined* directedness toward beings. Here we see why Heidegger wants to distance himself from the traditional understanding of nature: nature as the being of beings *just is* a predetermined way of being directed toward beings. The other inception, rather, begins by means of an attempt at understanding the historical transitions along the way of the first beginning and thereby shows the necessity of the questioning of being itself.

If the confrontation with the first beginning is not an opposition toward it, then what exactly is it such that it will lead to a nondominating relation to beings? The confrontation is a call for a decision, which leads us to the second point in need of clarification that will allow us to see Heidegger's resistance to the understanding of beings within machination. "Decision," for Heidegger, is another technical term. With it he designates something that happens through questioning of being, not an individual choice that one might make. When one engages in the questioning of the meaning of being, unconcealing machination as only one manner in which the being of beings can be interpreted and thereby questioning the very disclosure out of which correct statements are made, one is seeking the ground or basis of truth (*BP*, 102/70). This is absolutely not to say that one is seeking a certain foundation for either the discovery of or the construction of knowledge. Rather, since truth has its basis in the structure of unconcealment and concealment, since truth happens in the strife of earth and world, questioning being allows for a decision to happen, a decision that would remove humanity from its disastrous path through the recognition that the being of beings must be determined from within the essence of being (89/62). But what does it mean to say that beings must be determined out of the essence of being? Nothing more than the fact that the being of beings is determined within originary strife: a being is interpreted and given a place within a world on the basis of the relations that obtain between beings in light of some preunderstanding of what those beings are.

In order to see what this strife entails and its importance in turning away from machination, we can turn toward "The Origin of the Work of Art," where he defines earth and world not with respect to their opposition, but with respect to truth. Here, again, Heidegger defines *physis* in terms of an emerging, but adds that *physis* "lights up that upon which and in which humanity grounds its dwelling [*Wohnen*]" (UK, 31/21). He immediately goes on to say that this ground for the dwelling of humanity is earth. Earth, however, is not intended in the sense of our planet, but rather "that within which the emerging of everything that emerges returns for shelter [*zurückbirgt*] and, indeed, returns for shelter as one such emerging thing. The earth essences in that which emerges as that which shelters [*Bergende*]" (31/21). To pause here, there are several aspects of this rapid conceptual movement that require explication. First, the translation of *physis* is no longer "*natura*" or "nature," but rather "earth." This is part of Heidegger's attempt to think of the being of a being outside the metaphysical restriction of the essence of *physis* into one kind of *poiesis* among others and then on the basis of constant presence. Note, however, that his determination of the earth is not limited to his explanation of *physis* in *Introduction to Metaphysics*. He adds to it the fact that the *dwelling* of humanity, which is what defines the realm of the human for Heidegger, is based upon or grounded in the earth. This connection is significant because with it, Heidegger is able to make the connection between humanity, earth, world, and truth. "Dwelling" (*wohnen*) is etymologically connected to the verb "to habituate" (*gewöhnen*), to the commonplace (*Gewöhnlich*), and to habitude (*Gewohnheit*), and could also be translated as "to live" in the sense of inhabiting. What allows for dwelling, for habitual relations, for the commonplace, is world: earth is that which emerges, but necessarily emerges into a world. It is in this sense that earth is said to shelter (*bergen*) beings. The earth shelters insofar as it retains the mystery; it resists complete disclosure within the relations of the world, which are essentially open and subject to change. Earth embodies the open relations of the world, preserving them in a being and, as sheltering, remains something that closes itself (*sich verschließen*). World unconceals (*unverbergen*), while earth conceals (*verbergen*), but it is only together that a being is disclosed (*entbergen*) as a being. Thus, these two opposing movements belong together and are in strife. Just as for Heraclitus, what makes a being a being is the strife between earth and world. It is no coincidence, then, that Heidegger reintroduces Heraclitus's fragment 53 in order to explain the sheltering of earth (32/22): the earth shelters the relations of the world within a being. In that fragment, though, *polemos* "lets some appear as gods, and others men," and likewise in Heidegger's strife the valences of a world are sheltered in beings themselves. In other words, beings themselves reinforce or refuse certain ways they are disclosed within a world. The addition to the earlier account is that the strife, in making beings appear in a particular manner, also gives rise to "decisions" with respect to the human relationships with those beings. So, in the interrogation of beings and how they appear as beings, we have uncovered an originary strife between earth and world that gives rise to a determination of both human beings and the "gods."

A brief explanation of what Heidegger means here by "gods" should prove helpful. He is in no way invoking a belief in any supernatural power or being and views his notion of god as "outside those determinations meant by titles such as 'mono-theism,' 'pan-theism,' and 'a-theism'" because all "theisms" remain within the metaphysical tradition (*BP*, 411/289). "Gods," for Heidegger, are not beings at all, especially not beings outside the sensible world. Rather, a god is a "most extreme and briefest decision concerning what is highest" (406–7/286). It is in light of a god that beings take on worth within a world. For example, in the light of the Christian god, beings appear as hierarchically organized for human stewardship and use. Gods, that is, the way in which significance is distributed among beings, change, come, and depart through the strife of earth and world, which occurs within beings and in relation to humans.

Returning to the discussion of strife, we said that the strife of earth and world is that which allows a being to manifest itself as the being that it is. The final point to make concerning the strife in our context is that through the strife, through the manifestation of beings as something, certain relations are taken to be true. Following the Greeks, Heidegger once again begins by aligning truth with unconcealment (UK, 39/28), but, as he noted in Aristotle, unconcealment (*aletheia*) is equally a denial (*steresis*). In other words, whenever something appears to us to be true, this is due only to the ways in which we deny to the being other kinds of relations that are concealed to us through the specific mode of disclosure in place. In this essay, Heidegger speaks of concealment as a double phenomenon: as "refusal" and as "disguising [*Verstellen*]" (42–43/30). Concealment as refusal indicates the manner in which a being is not completely unconcealed and thereby remains mysterious. As disguising, concealment takes the form of a being manifesting as something other than what it is. These two forms of concealment correspond to the two elements in strife: earth conceals and holds within it relations that cannot be made apparent all at once, while world can make a being appear as something that it is not. Earth refuses to be made wholly present, and world sometimes disguises beings as what they are not. Because these concealments are always occurring, a being is unconcealed only along with certain denials. The essence of truth, then, requires both unconcealment *and* denial: truth and untruth belong together in an "originary strife" (43/31).

Rather than taking the interpretation of beings as absolute, then, Heidegger is taking pains to demonstrate how such interpretations are historical and therefore should be taken to be contingent, alterable, and co-constitutive of a given way of being in the world. This movement of co-constitution is precisely what is at stake in decision. On the one side, there is the preunderstanding of the being of beings that determines how those beings appear and therefore how one deals with beings. On the other side, one's dealings with beings affirm or contest that preunderstanding. What happens in the laws of machination is that this second side of the movement becomes increasingly reinforcing of the interpretation of beings, which leads one to believe that the interpretation is something more than a disclosing; it is seen as the one and only truth. With this double movement, we can see how questioning being would bring about a nonvolitional decision: questioning brings to the fore the manner in which a disclosure of beings underlies what is unconcealed and therefore also allows an acknowledgment of what is concealed and remains mysterious.

Once this structure of the decision is brought to the fore, it must be noted that there is no *one* decision that can occur concerning being. Through questioning the meaning of being, the possibility for decision is opened up, but the decision about being takes the form of myriad decisions with respect to one's relations with particular kinds of beings. The reason for this multiplication of decisions lies in the dual movement discussed above: if one's relations with beings affect the preunderstanding of them, then decision will take as many forms as there are kinds of beings with which to have relations. So, when Heidegger claims that all decisions are gathered into a single decision—"whether being definitively withdraws itself *or* whether this withdrawal as refusal becomes the first truth and other beginning of history" (*BP*, 91/63)—we must understand this in the context of the decision enacting a historical transition from the path of the first, Greek inception of philosophy to a path that does not send anywhere, precisely because it *refuses* to send. Through the refusal to send, the path opened lacks a destiny; it leads nowhere and has no goal, purpose, or end. The "refusal" spoken of in the decision refers, as earlier, to the manner in which the disclosing, which allows for the disclosure of beings, recedes into the relation itself and is concealed. In other words, in any singular instance of disclosure, the being appears *as* something on the basis of a fixed manner of disclosing that remains inconspicuous within the disclosure itself. The disclosing refuses to be unconcealed in the unconcealment of the being, and thus that which is denied of the being remains concealed. If the first beginning is marked by the loss of this denial in favor of defining truth strictly in terms of what is correct with respect to a revelation of beings, then Heidegger's challenge is to begin the other beginning with the recovery of this concealedness and an attempt to rethink the being of beings in light of being's refusal to be named and identified as a being. If taken as the "first truth" of the other beginning, the refusal creates an open relation to beings due to the fact that beings will retain their self-determination insofar as earth repels attempts to unconceal it rather than being mastered through a determination of the preunderstanding of them.

This openness takes the form of varied relations to beings that are significantly distinct from those of machination, each of which is embodied in a discrete change in relations to beings that Heidegger identifies as the decisions that follow from the decision. In *Contributions to Philosophy*, Heidegger enumerates eleven such "decisions" that arise out of being's being put up for decision, and it is telling that one of those decisions concerns nature. The decision, as Heidegger sees it, is as follows: "whether nature is debased to a realm for exploitation by means of calculation and ordering [*Einrichtens*], debased to an opportunity for 'lived-experience' *or* whether as it, as self-closing earth, bears the open of the pictureless [*bildlosen*] world" (*BP*, 91/63). The first option corresponds to what happens if being definitively withdraws, that is, if the path of metaphysics is not questioned and being is not put up for decision. In this case, the current situation with respect to nature will continue: the earth is a resource fit only for development, sustainable or otherwise, or for human enjoyment. Without questioning being, machination continues to disclose nature as a standing-reserve of energy. The alternative that Heidegger offers is somewhat cryptic as long as the "pictureless world" remains unclarified. Without going into too much detail, Heidegger has a discourse concerning the "world picture [*Weltbild*]" that largely consists of a structure similar to the one that we have already discussed with respect to disclosing above.[10] What is decisive, in the philosophical sense, about the world picture is that "beings as a whole are set in place as that whereupon man is established [*einrichtet*], as what he, therefore, correspondingly wills to bring and to have before himself and thereby, in a decisive sense, wills to place before himself [*vor sich stellen*]" (ZW, 82/67). In the age of the world picture, the world is placed before humanity in such a way that it is organized in a predetermined way that befits a particular use. What makes this possible is hidden in the last locution, "wills to place before himself," for the German word for "representational thinking" is *vorstellen*. Heidegger is bringing to our attention how the modern conception of a human being as a subject who places a picture of beings before itself so as to have knowledge of them—that is, represents beings and has representations that correspond to an external, objective world—also sanctions the subject's use of those "objects" so represented. The implication of the offered decision is that if nature is not to be taken as a "realm of exploitation," then there must be a transition away from the very idea of a human being as a representing subject.[11] We have thus come full circle to the question with which we began the last chapter, "Who are we?" The difference is that we can now see that the question must remain an open one that is answered through the ways in which "we" deal with beings because the "we" is determined only in relation to other beings.

There is something else, however, hidden in the call for a "pictureless world," that is, a world that is not a world picture: a call for a world that is not organized into a closed system or, as Latour put it, organized in advance and without discussion. If the world picture is a product of representing consciousness, of the attempt to establish the world as a plan that is laid out before humanity for its use, then "the being of beings is sought in the representedness of beings" (ZW, 82/68). Representedness is found within a system: beings have their place within the system that itself possesses goals and purposes. Heidegger specifies what he means by system here: "the unfolding, developing unity of structure within the represented as such, which arises from the projection of the objectness of beings" (ZW, 93/76). Heidegger finds fault in the construction of these systems precisely because of their closure. The closure of a system is the highest and most refined form of mastery precisely because it does not give the impression of being mastery: each being simply has its place within the system. The questions that Heidegger raises are: where the notion of a hierarchy of beings originated and why we think there must be such a hierarchy. Venturing an answer, he indicates Plato's hierarchy based upon a nearness to the ideas (*BP*, 273/192), asserting that the systems of German idealism are nothing but a new attempt at a Platonic staging of beings. Heidegger argues that, for Plato to be able to interpret the being of beings as *idea*, beings had to be experienced as *physis* for the reasons discussed above concerning the relation of *physis* and *eidos*. Aristotle then divides *physis* into the union of *morphe* and *hyle*, which plays out in terms of the form-content distinction in modernity. This form-content distinction is later found "at its strongest and most certain, and at the same time most rigid, in Hegel" (*BP*, 191/134).[12] In other words, Hegel's system is predicated upon the acceptance of the form-content distinction that is derived from the original determination of the being of beings as *physis*. But beings appear in stages only because being has always been conflated with the being of beings. Through the recognition of ontological difference and the subsequent attempt to respond to the grounding question of the meaning of being itself, Heidegger calls into question the very idea of this "staging" of beings. If, then, one were to think of being itself and not conflate it with the being of beings, the question arises whether beings could be organized into a system at all.

With respect to the idea of nature, this is a very poignant question. For, if there are only beings and no "thing" that is nature, and if one denies that there is no inherent order in the form of hierarchical stages among beings outside a given interpretation of beings, then it seems as if Heidegger is rejecting the very idea of something like nature. This interpretation does not seem to be an accurate assessment of his views, however, given his continued use of the word "nature" when describing the detrimental influence of machination upon beings within the environment. It is true that what we call nature, that ordered domain of beings that human activity did not create, no longer exists as a domain: earth is within every being, which is to say that nature exists only within beings. To say that a being is of the earth means that it emerges into a world of meaning in such a way that it also, as that which shelters the world, repulses every attempt to completely unconceal it: nature loves to refuse and disguise itself. If what one cares about is the preservation of the *environment* in a certain form, then it becomes necessary to find ways to alter the fundamental relations that one has toward the earth, ceasing the attempt to unconceal it definitively and totally. The challenge is not to find ways to leave nature alone, to eliminate human interaction with the so-called natural world, but rather to encourage forms of interaction, behaviors, and attitudes that do not dominate, challenge, and exploit the beings around us. What remains to be determined, then, is how Heidegger might allow one to conceive of nature after rejecting the possibility of a hierarchical systematization of nature. Though Heidegger does not himself address this question, such a possibility lies within the relational ontology that arises in his thought of *Ereignis*.[13]

II. *EREIGNIS* AND THE RESTORATION OF NATURE

Up to this point, we have largely been discussing Heidegger's negative views with respect to the philosophical tradition and its views of nature with minimal reference to his own positive view. The reason for this, as previously stated, is that Heidegger has very little to say about nature as such given that he considers the idea of nature as a substantial domain to be a modern development that is the product of a particular interpretation of beings. For Heidegger, like Latour, there are only beings and relations between them; there is no overarching system to organize them that exists independently of those relations. As a very particular kind of being, humanity has the potential to order and organize, but this potential is only a product of a more fundamental designation of humanity: Dasein. In this section, we will look to Dasein for ways in which those of us today who wish to continue to think about, speak of, and preserve nature might do so in a nonsystematic, nontotalizing, nondominating fashion.

For Heidegger, Dasein is a *transitional* determination of humanity, *not* the definitive determination of its essence. Dasein is the determination of humanity that must be achieved through an asking of the grounding question in order for the other beginning of history to come about, *not* another name for a human being. Dasein is something that happens, *not* itself a being. Dasein is the "projecting open of the truth of being ('there' [*Da*])" (*BP*, 295/208) and, as a projecting open, is an open structure. These preliminary determinations are rather cryptic and require further elucidation. What we are meant to take from them at this point is that Dasein is something that is possible within the first beginning, but at the same time leads toward something other than what is destined out of that beginning. As such a transition, Dasein itself is a *crisis* between these two inceptions of thought (295/208) and, as a crisis, will have a double determination that comes from its position of having a foothold in both beginnings. From within the first beginning, Dasein refers to that which is actual, real, and existing. This definition of Dasein corresponds to the determination of truth as unconcealment and unconcealment alone. Since what is unconcealed in the first beginning is unconcealed to either the senses or the mind, what exists is determined as that which is either sensed or thought. This limitation of the possibilities for conceiving of existence led to the opposition between so-called "realism" and "idealism." But the notion of truth conceived as total unconcealment is derived from a grounding in the being of beings, which is a response to the guiding question of metaphysics. As we have held from the very beginning, Heidegger is pursuing the deeper, grounding question of *how* beings appear as the beings they are: "How does being essence?" (387/270). In other words, he wants to discuss on what grounds one can respond to the guiding question, What is the ground of the assumed foundation? This question leads to another sense of Dasein that is not opposed to the first but rather is that which gives the first its particular sense (297/210). This sense of Dasein originates in the perspective of the not yet attained other beginning of thought that takes Dasein to be "the being of the there [*das Sein des Da*]." Heidegger goes on to define the being of the there as "the openness of a being as such in the whole, the ground of the more originarily thought *aletheia*" (296/209). Thus, from within the attempt to begin anew, Dasein attains a distinct sense, but one that is itself not a *fixed* determination precisely because it is defined in terms of the openness to beings as a whole, that is, an openness to the being of beings. But additionally, Dasein is what grounds *aletheia*, which, as we saw in "The Origin of the Work of Art," is produced through the strife of earth and world. So, in making the connection between Dasein and the grounding of truth, the "*Da*" becomes the site within which the strife happens, and strife occurs in the clearing. Dasein is, for that reason, that which has the ability to bear and sustain the clearing, "die inständige Ertragsamkeit der Lichtung" (298/210).

Dasein, as the transitional point between the first and other beginning, loses its sense as what is "in being" and takes on the new sense of that within which being first essences (*west*) or comes to presence (*anwest*). As such, Dasein must be questioned so as to see exactly *how* being comes to presence within a being. Heidegger's word for being at this transitional stage is *Ereignis*, the happening of the strife within the there, the projecting open of being. The strife between earth and world gives rise to truth, with the world disclosing beings in a particular manner and the earth preserving those relations so as to conceal others. In the last chapter, we described how the disclosure of the world places humanity upon a particular path and called this being "on the way" and the transitions between paths and within a path "history." Consequently, if *Ereignis* is the happening (*geschehen*) of the strife, then it is originary history (*Geschichte*) as well. In this sense, truth is a historical phenomenon. *Ereignis* is also the name for how being essences (*BP*, 30/22). We must be cautious in how we understand this statement so as to hear it properly. That being essences as *Ereignis* does not mean that being, as something that exists, takes on form through the strife of earth and world or that it appears among beings because being never exists as a being. Rather, *Ereignis*, being, essences; and its essencing is "the settlement [*Austrag*] of the strife and objection" (30/22). Taking *Ereignis* as a "settlement," as the resolution of a struggle, explains the constancy and continuity with which beings appear within the world, but we run the risk of misinterpreting *Ereignis* if we do not also note the transition from the verbal *wesen* to the substantive *Wesung* that takes place in the determination of being as something resolved. Within being, that is, in the constancy of being's essencing, beings are: they appear as beings, and they do so as if they are the beings that they are outside all human relations to them. Being, however, is not a being; it essences: it is that which allows for the being of beings to be determined (correctness) through a projecting open of the world. Hence, *Ereignis* occurs at every moment: it is constantly in the process of resolving the strife, but the strife is continually renewed. So, when Heidegger speaks of the "uniqueness" of being, he is referring to the singular manner in which being is projected open at a given point in history. Truth is historical because being itself is historical.[14]

Dasein, however, is the ground of truth because it is within the *Da*, within the clearing, that truth as unconcealment-concealment arises. Accordingly, Dasein *ereignet*;[15] Dasein projects open a world while at the same time dwelling within the earth. Heidegger expresses this alternatively by calling Dasein the "thrown thrower" (*BP*, 304/214), meaning that just because Dasein is that which projects the world open does not mean that the projected world originates or is created by Dasein. Dasein is not the author of the world and, as a being in the world, is not the author of itself. In the projecting open, beings appear as beings, which implies a response to the question of the being of beings, the mark of representational thought. The representation covers over the original projecting open, hence the reason why Heraclitus claimed that being (*physis*) withdraws itself. The laws of machination within which the enframing of beings recedes into the background and becomes inconspicuous, then, is a more specific incidence or characteristic of representation in general. For representational thought in general, there is a desolation of being (*Seinsverlassenheit*) among beings due to the way in which the representation prevents the projecting open from coming into view. If, however, Dasein is taken as the clearing within which beings appear as beings, then the projecting open becomes questionable for the reason that it is no longer concealed within the representing. But Heidegger's goal is not to explain how representing happens, explain the projecting open, but rather to alter the manner in which the projecting open assumes a response to the guiding question and in so doing challenge representational thought in general (*BP*, 326/229). If representation encloses a being in a given signification, the projecting open that Heidegger envisions would contrarily "let the being be," which is to say not to dominate them in a representation.[16]

At this point, we can begin to discuss how these notions of Dasein and *Ereignis* might be helpful in our thinking about nature. In order to make the connection, however, between the stated project of opening up the possibility for another beginning and current thinking about nature, there is one last obstacle to overcome. Fortunately, it is one that Heidegger also addresses and to which we now turn. Nature, no matter what else, is thought of as a spatial phenomenon: nature is a specific place, if not *the* place. In physics, nature is the spatial-temporal whole: everything that *is* is within time and space. If one accepts Heidegger's arguments with respect to the history of the Western world, however, a problem arises for this conception of nature insofar as it is equated with the conception of Dasein from the first beginning. For classical physics, everything that exists does so within a pregiven spatial and temporal matrix. Hence the difficult cosmological questions that arise within physics itself with respect to the finitude or infinitude of the universe, the number of universes, and the current mystery surrounding the presumed existence of "dark energy." But it is germane here to note that these problems arise from *within* modern physics, which indicates that they are questions that come about due to the playing forth of the grounding premise of the science. Were that premise altered, or were the science itself to become philosophical and question its own foundations, these questions would not, perhaps, arise. Currently, a question that is arising within physics is whether the concept of gravity must be revised so that the empirical data achieved through experimentation can be accounted for in the physical representation of the universe. Of course, this amounts to a reordering of the representation of the system, not to a displacement of the primacy of a representable system.

We should understand Heidegger's thinking about space and time, then, in the light of the project of questioning the fundamental premise of physics, not in terms of an attempted counteraccount of the structure of the cosmos. If Heidegger were endeavoring such a counteraccount, he would merely offer another representation of the world rather than capturing exactly how this world is thrown open from Dasein. He would then fall prey to his own criticism of representation and, in the end, have nothing to offer in terms of a new view of nature.[17] But in challenging the very foundations of physics, and thereby of all the sciences, Heidegger does not attempt to supply another grounding premise so as to close the region of inquiry once again and allow for the science, an institution (*Einrichtung*), to play forth once more in a different direction (*Richtung*).

One cannot, however, simply say that the notions of space and time that ground physics are in need of revision and move on. He must also justify precisely why the ordinary (*gewöhnliche*) representations are not sufficient. To this end, Heidegger supplies a number of statements at the beginning of *Contributions to Philosophy* that indicate the course of the investigation, some of which have already been discussed (*BP*, 30/22). Here, he designates the importance of "time-space" as the "fundamental framework [*Grundgefüge*] of history" that itself arises from out of that history. Time-space is the fundamental framework of history because it is the "site for the moment [*Augenblicksstätte*]," a determination that recalls the prior claim that being is *continuously* being projected open *at each moment*. The language here is not coincidental. In *Being and Time*, "the moment" is Dasein's self-temporalizing, which is to say how Dasein first experiences itself as "in time" in the ordinary sense at all (*SZ*, 328/302, 338/311). The moment is more fundamental than the conception of time as a linear series and is in fact what makes linear time possible. How is this so? Because Dasein experiences itself within time on the basis of projecting forward particular projects for itself and finding itself within a particular situation from which it must attempt to achieve those projects. This structure is called "care," and if Dasein experiences itself in this way, with the past and future constantly encroaching upon it at all times, the moment will, in a certain sense, elongate itself such that the origin of the projection is no longer thought of as having its source in Dasein. Time then becomes posited as a property of the world: the not-yet becomes the future, and the already-been becomes the past, which are both to be distinguished from, and as no longer immanent to, the present. The separation of the not-yet and the already-been within the moment then transforms the moment into the instantaneous present (*Gegenwärt*).[18] Because of the temporalizing of Dasein, it also finds itself "factically entangled" (*SZ*, 368/336): it has a location among beings that is determined through the structure of care. Location is the spatializing of Dasein through directionality (orienting itself in terms of other beings) and de-distancing (*Ent-fernung*) (the making use of or interacting with beings that are at a distance in a given direction from Dasein). Space is not *dependent* upon time, but the neglect of the relation of time and space is one of the primary sources for the rise of the primarily spatial representation of nature (369/337–38). Though Heidegger does not say as much here, once nature is divorced from time, it can then be asserted that it is in fact outside history. If, however, the relation between space and time is maintained, it makes much more sense to speak of nature as a historical phenomenon, but it is important not to confuse the sense in which nature as a spatial domain is also a temporal one with the ordinary representations.

In order to accomplish such an analysis of space and time in their original unity and complete our picture of nature, we must ask how spatializing and temporalizing are unified such that a cohesive picture of nature can arise out of them. Heidegger offers such a picture in *Contributions to Philosophy*. In principle, the same is the case in this text as in *Being and Time*, but rather than arriving at his conclusions via an "analytic of Dasein," where Dasein is taken to be the kind of being that human beings are, Heidegger is looking more directly at the manner of projection itself rather than the contents of said projection. As a result, the analysis is much richer, but also more abstract insofar as he looks to how notions of time and space arise from out of a projecting open of being from within the structure of the *Da* and not from out of human experience. This is part of the move away from anthropocentrism described above. In the same way that *Being and Time* provides a description of how Dasein is temporalizing and spatiality is grounded in this temporality, time-space, as the site for the moment, is prior to either space or time and should not be thought of in the sense of modern physics, where "space-time" is taken as a single continuum within which time is a consequence of a spatial mass and vice versa. Time-space, as the fundamental framework of history, is also a historical notion.

If this last claim seems a bit circular in its logic due to the inability to discern which term grounds the other, this, once again, is not coincidental. To explain briefly, let us begin with elements already in play. Truth is defined in terms of the structure of unconcealment-concealment: for any unconcealment, other ways of the being's manifestation are concealed. Beings are unconcealed, however, through a mode of disclosure that provides a ground for the correct or incorrect statements about an object. These disclosures that provide a ground are within a given world that is directed toward the earth that shelters beings from the disclosure of world. The strife occurs within the *Da* and inaugurates (*eröffnen*) a historical moment. But throughout all this we must note that the ground for truth is mutable because truth is relational, and there is nothing that requires any specific kind of relation as the ground of truth. Hence, truth is grounded upon an abyss (*Abgrund*): contra Descartes, one does not find certainty in the foundation for knowledge because the foundation itself only arises from within a world, which is itself groundless. History, then, consists in the transitions in ground, which, in this context, we can say are the various ways in which the abyss grounds truth. Every grounding of truth, that is, every resolution of the strife of earth and world, is historical, and we called this resolution *Ereignis*. But Dasein is that which *ereignet*, and so, here within the *Da*, truth is grounded in a historical manner by means of the temporalization of Dasein. Temporalization and spatialization, however, always come together, and so Dasein, as the ground of truth and hence of the abyss, is also found to be the ground from which beings are temporalized and spatialized such that they then may become ontic time and space.

Nature, which is what we ordinarily refer to as the objective domain of space and time, is for Heidegger a completely derivative phenomenon. The fact that time and space are derived from more fundamental temporalizing and spatializing processes within Dasein does not, however, reduce them to a mere human projection. Rather, space and time as we experience them are relations as well. As relations, space and time find their source in time-space, the abyss of the *Da*, but how does this work? A being is the being that it is through a projecting open of being and, in the case of the first beginning, a projecting open of a being as that which is constantly present. Constancy is a temporal notion, while presence is a presence in space. Thus, in the first beginning, the ground of truth as what is unconcealed also requires a definitive interpretation of space and time. But when the projecting open is questioned, those notions of space and time must also be redefined in order to accord with the groundlessness of truth. Time thereby becomes a "rapture (*Entrückung*) and thus an inauguration" (*BP*, 191–92/134). "Rapture" not only means "captivation" here, but has the additional sense, coming from its relation to the verb *entrücken*, of being carried away. Heidegger is calling our attention to the manner in which we become absorbed into time, and, being so absorbed, we are carried away and hence inaugurated into a particular mode of relation with beings. Time is a way of being related to beings. Part of this rapt inauguration is the perception of beings *at a distance*, which, as mentioned above, is how the time-relation spatializes (*einraumen*). So space, when taken together with time as that which inaugurates one into a projecting open of being, is not an objective domain or container of beings, but rather is "spacing [*Räumung*, more literally 'evacuation' or 'removal']" (192/134). Because spacing occurs through the rapture of Dasein in time, space is able to charm or enchant (*berücken*) Dasein into the belief that it has always been there. Presence (*Gegenwart*), then, is the disguising (*verstellen*) of the temporal nature of beings by means of duration's (*Dauer*) seeming to be mere perseverance (*Ausdauern*), and, in so doing, spatial presence seems to be timeless (192/134). Time and space always come together; time-space projects open beings and, as spatializing and temporalizing, "constitutes the essence of beingness, constitutes intersecting" (192/135). Since Heidegger's thinking about nature has shifted to a thinking about individual beings in their relatedness to other beings in time and space, time-space, as the essence of beingness, is that out of which a new thought of nature must arise.

In order to stipulate what nature might look like within this other beginning that takes *Ereignis* as its point of departure rather than a certain determination of the being of beings, we must inquire into the rapture and enchantment of Dasein to see how this will provide *another ground* rather than mere relativity. In the projecting open of being, we said that the abyss is that which provides a ground for the encounter with beings. The ground that arises from the abyss, however, can never provide certainty because certainty comes about by confusing the projecting open for something more than an abyssal ground, falling into representational thinking. The ground is such that it will not allow total unconcealment of a being in its essence, but rather preserves the mystery within beings by refusing access to all possible unconcealments in a given disclosure. Because the abyss both gives a ground but refuses for it to be unshakable, refuses a solid ground, Heidegger calls the abyss "the deferring refusal of ground" (*BP*, 380/265). In this deferral, there is not only the refusal of total unconcealment, but also a clearing.

We must digress in order to note the implication of the definition of the abyss as deferring refusal: if the ground is constantly deferred and refused, then *being is always up for decision*. The decision about being is not a decision at a single point in ontic space and ontic time that, once made, will never have to be revisited. If a decision were final, it would amount to another sending into representation. Rather, the decision is one constantly being made: one must hold oneself within the moment of decision, pause within the undecidability of the moment so that a god may be inaugurated (*BP*, 23–24/17, 382/267). Dasein, as the projecting open of being that is itself projected, as the thrown thrower, is always able to put the established order and worth of beings up for decision, and indeed must consistently do so if Dasein is not to dominate beings. In this continual dwelling within the decision, by dwelling in the undecidability of gods made possible through a recognition of the abyss as ground, humanity loses its place as the master and orderer of beings, opening the way toward another kind of mastery in which thinking is held open. The import of this last idea will become more relevant once the inquiry into time-space is concluded.

To return, then, to our previous discussion, we will shift our focus from the clearing of beings, which we have been emphasizing throughout, toward what is relevant to us here: how the deferring self-refusal is already a temporalizing and spatializing notion. Grounding occurs through the temporalizing and spatializing of beings, but they are meant to be understood *absolutely* in the Hegelian sense of being both a unity (*Einheit*) *and* maintaining their separateness (*Geschiedenheit*) (*BP*, 383/267). We already called temporalizing a rapture and spatializing an enchanting, but we did not inquire into how they comprise the openness to beings. Heidegger gives us four statements that are meant to show the connectedness of the two while at the same time maintaining their distinctness. They run as follows:

> Space is the abyssal grounding that enchants of the surrounding support.
>
> Time is the abyssal grounding that enraptures of the gathering.

> The enchanting is the abyssal surrounding support of the gathering.
>
> The rapture is the abyssal gathering into the surrounding support. (*BP*, 385/269)[19]

The four operant terms within these statements are completely intertwined, but in their interconnectedness show themselves as distinct. The dual movements of temporalizing and spatializing are most clearly seen in the latter two statements. The carrying outside oneself (rapture) of time gathers beings into a surrounding support: time forms a world by making Dasein ek-static, but the basis (*Grund*) for this world is the abyss (*Abgrund*). The enchanting of space is the absorption within that which is gathered: Dasein is always supported by a world, but that world's support is abyssal. The reason that the surrounding support is abyssal is that the gathering occurs only through time. Since rapture is the primary action of time thought from within the questioning of the projecting open of being, the gathering is contingent upon the projecting open. Defined in a similar manner, space, whose primary action is enchanting, is what is gathered together and thereby is able to provide a foothold from which Dasein can dwell among beings. This foothold is, however, tenuous in its arising from the abyss. Without time, there is no gathering that will provide a foothold, but without space, there is no support to rely upon that will hold the world together. The mutual dependence of time and space stems from their common source in time-space, that is, the abyss.

Time and space, however, are not a simple unity. As mentioned above, they also must maintain their separateness. To explain how this is so after placing them in such close interwoven relation, Heidegger offers two more statements (*BP*, 386/269):

> Time spatializes (*räumt ein*, also "grants" or "allows"), never enchants.
>
> Space temporalizes (*zeitigt ein*, also "brings about" or "leads to"), never raptures.

Time is that which creates possibilities within the world by opening up the world in a given way, but it is not capable of absorbing a being within that world; space is that which enchants. Space, on the other hand, concretizes those possibilities, makes them manifest within beings by preserving worldly relations in the earth, but is not capable of carrying a being outside itself; time is that which raptures. We must be cautious not to reinvest the possibility-actuality distinction in interpreting these sentences, however, because Heidegger does not want to be interpreted as merely restating the metaphysical system of modality.[20] In considering modality, there seems to be a mode of being that is possibility, another that is actuality, and a third that is necessity; all three modes of being are defined in advance in terms of how being is projected open. If the projecting itself is taken as abyssal, that is, not taken as fixed, then possibility, actuality, and necessity are all to be created in the decision about being.

Time opens space, and space holds open time, but only in their interaction do either one of them essence as that which it is. The play between the two mirrors the strife between earth and world, with time as the opening of world and space and the sheltering of the earth. Just as the truth produced through the strife is preserved in a being, so, too, is time-space preserved within a being because it is in the fixing of a being as a being that time and space are differentiated and become explicitly manifest (*BP*, 382/267). Beings do not manifest within space and time; space and time manifest as they do because being differentiates itself from beings by temporalizing and spatializing beings. Time and space are relations between beings as they enter into the openness of the *Da* and as such are preserved within beings, not the reverse. If the problem with the concept of nature was mistaking it for an objective realm of space and time, then any expectation of preventing the domination of nature does not lie in seeking its intrinsic value in objective properties, but in looking to the manner in which beings preserve themselves in relation to one another.

III. CONCLUSION

With the idea that time and space are relations that can be discovered within a being, we arrive at a point where the idea of nature can be restored. The criticism of nature revolved around its being defined in terms of time and space as objectively existing dimensions that allow for the organization of nature into hierarchical stages. Taking this one projecting open of being as absolute or as real allows for exploitative dealings with other beings insofar as they are reduced to only that which is unconcealed within a very particular understanding of beings as constantly present. Nature, in fact, embodies the representation of space and time as objective that arises within the first beginning. This idea of nature itself, on Heidegger's account, was responsible for sanctioning the devastation of the earth.

In order to stop that devastation, the idea of nature had to be removed from its privileged place in order to make perceptible the decision that must be made with respect to nature: Is nature just something there to be used and ordered, a calculable realm of objective things? Or is nature thought of as earth, that which bears the world? Each of these alternatives corresponds to a more fundamental decision with respect to nature: Will being definitively withdraw from view and allow the complete domination of both nature and humanity, or will the withdrawal of being be recognized as the preservation of the earth, the refusal of total disclosure, of ordering in terms of human purposes alone? Questioning the meaning of being displaces the centrality of nature as the object that provides a foundation for all knowledge, truth, and ethics, while *at the same time* opening up the possibility that nature, thought in a different manner, might assert itself and resist human domination. This displacement takes place in the recognition that there is no *thing* that is nature: nature is preserved within beings in the way the projecting open of beings allows them to manifest. By showing how it is possible to question the organization of nature into systems, Heidegger opens up a way of thinking about nature that cannot rely upon the science of ecology, which, as Evernden has convincingly argued, can support pictures of nature as either competitively or harmoniously organized due to ecology's need to formulate itself as a science based upon the idea of nature as a system.[21]

Especially if ecology is going to be rejected as offering a guiding *normative* view as to the "nature" of nature, there remains an open question within Heidegger's thought with respect to nature: "how to order the living, 'nature,' and the lifeless within nature such as tools, machination, work, deed, sacrifice and their force of truth" (*BP*, 274/193). After all, even within Heidegger's examination of the projecting open of being, *Ereignis*, there remains a god, something that is highest, and so the world is ordered in a particular way. It has *not* been argued in this chapter that Heidegger is searching for a way to avoid the positing of all order in the world. Here, too, he is quite close to Latour insofar as *order itself* is not the enemy, but a certain understanding of what the order is and where it originates. Without the perception of some kind of order within the world, it would be impossible to function therein. The key is not to allow that order to become hierarchical through its absoluteness, to grasp the projecting open as historical. Accordingly, the argument has been that, through the recognition of the ways in which order is a historically contingent phenomenon projected upon beings and is therefore not an intrinsic property of beings, a different kind of nonrepresenting, non-dominating relation with beings then becomes possible.[22] Since Heidegger's work is avowedly preparatory for such a relationship, it is difficult to articulate what such a relation might be concretely within his philosophy. And so Heidegger leaves us with a question, a question to which he provides only the preparation for a response:

> Does there nevertheless not remain a way, at least provisionally, to create, as with the type of "ontologies" of diverse "regions" (nature, history), a prospect[23] [*Gesichtskreis*] for the projecting open that is consonant with being [*seinsmäßigen Entwurf*] and thus [to make] the regions experienceable anew? Something like that can become necessary as *transitional*; yet, it remains awkward [*verfänglich*], insofar as the slip into a systematization of the earlier style easily suggests itself from that prospect. (*BP*, 274/193)

Such is the challenge that faces us in our thinking about nature in a new way. We can treat it as a region as it used to be, a region of space and time, but we must exercise an extreme caution that we do not slide back into an idealistic system-building regarding nature. In order to avoid this backsliding, we must think of nature in terms of the time and space that emerge within beings from time-space, the abyss, that grounds truth. With Heidegger, then, we can still be "naturalists" and not reduce nature to a human projection, but nature can no longer mean a substantial realm of time and space, can no longer be normative either ethically or epistemologically; in short, can bear no resemblance to the nature that supports naturalism within contemporary philosophical discussion. From the perspective of a rethought nature that aims at an alleviation of the domination of beings, we might then begin to ask how best to live as part of nature, how best to dwell upon the earth.

But Heidegger does not construct this conception of nature for us. He leaves us with a suggestion of how to turn away from the metaphysics of nature toward a nature that is not enclosed and dominated by human activity, but is a partner to it. In order to construct the prospect from which to view this restored nature, we will now turn our attention away from Heidegger and toward the philosophy of Merleau-Ponty for the reason that his philosophy arguably offers us the richest articulation of a contingently ordered, but not systematized, vision of nature within the phenomenological tradition.

CHAPTER 4

MERLEAU-PONTY AND NATURE AS THE COMMON WORLD

> If an ecophenomenology could give us better access to nature than that represented by the naturalism which phenomenology was created to resist, by supplementing intentionality structurally with non- or pre-intentional characteristics of nature, would not ecophenomenology be the future of a phenomenology, one which has purged itself of its opposition to nature?
>
> —David Wood, "What Is Ecophenomenology?"

> Everything said about the sensed body pertains to the whole [*entier*] of the sensible of which it is a part, and to the world.
>
> —Maurice Merleau-Ponty, *The Visible and the Invisible*

From Merleau-Ponty's very first articulation of his philosophy, the central question of his thought has been the human relationship with nature, and from the very beginning he has opposed the conception of nature that has been argued against throughout this book. Merleau-Ponty's philosophy is marked by the attempt to see in every formalization of experience some grain of truth that inspired that particular philosophical expression. This tendency can be seen at the very onset of *The Structure of Behavior*, where he introduces the traditional definitions of nature from the materialist and intellectualist traditions in order to, over the course of the book, reject both and take from each the intuition that inspired it. Thus, nature is neither "a multiplicity of events external to each other and bound together by relations of causality," nor is it "the ensemble of objective relations borne by consciousness."[1] At the same time, though, Merleau-Ponty will retain the intuition from naturalism that human consciousness does not shape the world according to its immanent categories and from the critical tradition the insight that there can be no direct and unproblematic access to that world without thinking through relations. Though I will argue that this early stage of Merleau-Ponty's philosophy remains somewhat indebted to the critical tradition he criticizes, as his philosophy developed he increasingly distanced himself from the critical tradition. As is well known, the starting point for Merleau-Ponty's philosophy is the body's *involvement* in a world, beginning with the brute, prethetic contact of body and world. In trying to construct a prospect of nature from within this philosophy, what will concern us most throughout this chapter is not the nature of the body or the body's role in instituting meaningful relations with the world, but rather how to tie together Merleau-Ponty's perceptual philosophy with his attempted ontology of nature. Thus, the project of this final chapter is to describe the manner in which Merleau-Ponty's philosophy can be viewed as an attempt to create the prospect of which Heidegger spoke in *Contributions to Philosophy* from which nature and history can be experienced anew.

David Abram has already elaborated one such attempt at renewing the human relationship with nature through a phenomenological philosophy based in perception in his *The Spell of the Sensuous.*[2] In that text, Abram pursues what he rightly sees as the implications of Merleau-Ponty's philosophy, particularly those elements that pertain to language and expression, for the relation of humans to the "more-than-human world." *The Spell of the Sensuous* is the expansion of his earlier assertions concerning the undeveloped ecological consciousness that he sees at the heart of Merleau-Ponty's thought.[3] Abram argues that humans have become cut off from the natural world that sustains their lives and advocates a "reinhabitation" of nature by becoming more closely attuned to the natural processes of the land that surrounds us. The first step toward addressing the devastation of the earth that humanity currently perpetrates is to renew the lost connection that humanity feels toward the land, "to begin to recall and to reestablish the rootedness of human awareness in the larger ecology" (*SS*, 261). He argues that the Western transition from an oral tradition to a tradition based in the "technology" of the alphabet enabled a break with the sensual existence of human beings and a disparagement of that sensorial existence (cf. 93–135). The solution to this alphabetic distancing is not to abandon writing, but to rediscover in the living landscape the referents of our language. He does not intend to say that language is a set of correspondences between words and objects, but that language expresses relationships between different beings. If we understand language as articulating the world in this way, we begin to see the earth itself as speaking to us, not literally in human language, but as full of latent signification. Addressing the destruction of the natural world begins with "direct, sensory interactions with the land around us" so as to be able to "appropriately notice and respond to the immediate needs of the living world" (268). The written word can then be deployed in efficacious ways to transform our relations with the land because it will have rediscovered our relation of "reciprocity with the myriad things" (270). Abram also embarks upon a prolonged description of these relations in terms of the body's space and time as found in both Heidegger and Merleau-Ponty, drawing the conclusion that each of our individual perceptions of space and time is "granted to us only by the earth" (216). Discourses about the earth, then, should be place-centered, not only addressing the particularities of the landscape at hand, but also with modes of expression inspired by those places.

While meritorious in many ways for anticipating a great deal of Merleau-Ponty's thinking with respect to nature *prior to* the publication of Merleau-Ponty's course notes on the subject, we can develop Abram's line of thinking in different directions and in further detail given the material found in those notes.[4] Of course, given Abram's professed emphasis upon reinvigorating expressive, creative storytelling as a mode of knowing nature, it is clear that he did not intend a strictly academic audience for his book, and so this lack could hardly be considered a fault of his work. On the whole, I agree with Abram that there is a sense in which we are called to go beyond Merleau-Ponty in order to realize the true ecological value of his thinking, and, like Abram, I will argue that this can be accomplished from within the resources of his own thought. In specific, this transformation entails developing the notion of "flesh" in terms of a generalized embodiment, rather than being a specific aspect of human being or human perception, which is how it is usually interpreted.[5] On the basis of this reworked conception of flesh, Merleau-Ponty is able to provide a conception of nature that does not fall prey to Latour's and Heidegger's criticisms of the concept. The consequences of this association of nature and flesh are far-reaching and will involve describing in greater detail what Abram calls "the ecology of language" (*SS*, 77), so as to establish more strongly the interconnection and inseparability of nature and culture.

In the attempt to address Merleau-Ponty's thoughts concerning nature in this manner, however, we encounter a difficulty that requires us to respond to a question Renaud Barbaras poses in his essay "Merleau-Ponty et la nature."[6] In that essay, Barbaras first shows in detail why Merleau-Ponty would begin to give nature such a central role in his philosophy, provided the problems that remain from his earliest works, *The Structure of Behavior* and *Phenomenology of Perception*, and then proceeds to offer a structural outline of the later conception of nature.[7] The transition in Merleau-Ponty's views on nature, according to Barbaras, can be attributed to the larger transition from a phenomenological methodology to ontology, which is a transition we tracked in the last two chapters in Heidegger's move away from lived-experience. Barbaras concludes his line of argumentation with the question of whether the philosophy of nature that emerges from Merleau-Ponty's later philosophy ought to be viewed as an abandonment of phenomenology or the fulfillment of its highest demand. Because a traditional phenomenological standpoint explicitly orients Abram's project, responding to Barbaras's question is important in order to see how the relations between beings are conceived of differently depending upon the conception of phenomenology at work. In other words, if we answer with Barbaras that Merleau-Ponty's philosophy of nature does not renounce a phenomenological approach but fulfills its highest demand, it remains for us to determine to what extent this response alters Abram's account of nature due to *The Spell of the Sensuous*'s emphasis on experience. As we will see, Merleau-Ponty follows Heidegger in the rejection of lived-experience, and, consequently, rather than thinking about the human relationship to nature as one that is personally experienced and lived immediately, as Abram would have it, we must instead turn toward a different kind of experience of nature, an instituted, and thus mediated, experience. In this way, Merleau-Ponty's philosophy is oriented by the same concerns Latour will later describe in *We Have Never Been Modern*.

With the introduction of the idea of an instituted experience, we arrive at the main difficulty that this line of interpretation faces: If Merleau-Ponty rejects the idea of direct experience and views all experience as instituted, how, then, is it possible for him to define nature as "the primordial, that is, the non-constructed, the non-instituted"?[8] As Barbaras emphasizes throughout his essay, the definition of nature as the noninstituted arises because Merleau-Ponty's philosophy begins with the body's perceptual contact with "brute" or "wild" being: nature is that which is encountered, and thus "through nature, the sense of the being of *natural being*, the new name for the perceived, is what is in question."[9] If phenomenology begins with what is given in experience, and experience is given through various institutions such as perception, language, personal experience, and history, then there can be no direct experience of nature. Yet, in order to avoid the Kantian bifurcation of the thing in itself and the phenomenal object, Merleau-Ponty will maintain an *indirect* access to nature through instituted structures. In this view, nature exists independently of human cognition, but still remains the soil of all living existence.

The aim of this chapter is to arrive at a conception of nature that is neither mechanistic in function nor teleologically organized into a system through the resolution of the themes mentioned above. In order to do so, we will begin by looking briefly at Merleau-Ponty's early views on nature where he tries to deconstruct the nature-mind opposition by means of a reformulated theory of behavior. This starting point will prove integral to our future investigations by presenting the centrality of behavior and form (*Gestalt*) as mediating concepts that are meant to establish a third way of conceiving of nature. From there, we will look at how the notion of behavior transforms when it is generalized into the structure of expression. Expression, perhaps the central structure in Merleau-Ponty's philosophy, introduces mediation and institution as a priori structures for *all* interaction between humans and other beings. The next step in our inquiry will be to remove this structure of expression from the human realm and generalize it to all beings. This aforementioned generalization of Merleau-Ponty's notion of flesh allows us to see the myriad ways in which nature is the common world whose sense is derived from the collective expression of existing beings. Put in Latour's terms, and as we will see Merleau-Ponty uses the same phrase himself, nature will be the common world all beings dwell within. In the last phase of the argument, we will evaluate how these diverse ontological insights contribute to a new view toward nature. In this endeavor, we will be helped by looking at Merleau-Ponty's endorsement of Alfred North Whitehead's process view of nature. In doing so, we will discover a nontotalizing view of nature that, intertwined with history, will allow us to resolve some of the more prevalent confusions of environmental philosophy.

I. THE BEHAVIOR OF NATURE

The main problem that Merleau-Ponty seeks to address in his early work on nature is that of the limitation of the possibilities for thinking about the organization of nature into either a mechanistic whole—one whose identity lies in the aggregation of its individual parts—or a vital whole—one whose identity lies outside the natural processes themselves. In *The Structure of Behavior*, Merleau-Ponty suggests that the idea of behavior will allow science to overcome the impasse between those who see life as nothing more than a fixed set of physiochemical phenomena and those who see life as something essentially distinct or emergent from physiochemical phenomena. What the examination of behavior yields that is so novel is that neither account of natural existence can explain behavior without either leaving certain phenomena unaccounted for or giving rise to certain paradoxes that are insurmountable from within the presuppositions of the original metaphysics. To describe these arguments in sufficient detail would lead us away from our current questions, however, and so we will focus simply upon the ventured solution to the original antinomy structure. In the notion of structure, Merleau-Ponty believes that he has discovered a way of resolving the antinomy between mechanism and teleology in the same manner that Kant resolved the three antinomies in the *Critique of Pure Reason*, by uncovering the prethetic ground upon which each alternative is constructed and showing why the two alternatives are themselves asking their questions in the wrong way. Vitalism and mechanism share the common assumption of "realism," meaning that they both take everything that occurs and exists to be a cause of some sort. This version of realism is in turn possible only on the basis of a substance-based metaphysics, which has been the object of criticism throughout this book. Merleau-Ponty's early hope is that a formalist metaphysics, that is, an ontology based on relations that accepts structures as real, but not existing as things, will substitute for the metaphysics of substance (*SC*, 142–43/132). The move away from a substance-based ontology will undermine the antinomy between mechanism and teleology, between traditional forms of materialism and intellectualism, by showing how there are "real" elements to our experience that are not constantly present in the manner of a thing. In this tendency to redefine realism and what it means to be a material body, Merleau-Ponty has much in common with contemporary materialists inspired by Deleuze—such as Isabelle Stengers, Jane Bennett, and Latour himself.

In his early work, Merleau-Ponty relates form and structure quite closely to the concept of system. Form exists wherever there is a system that meets two conditions: one, that a change in any of the constituent parts alters the properties of the system as a whole; and, two, that if there is a change in all the parts of the system, then the proportions of the relations between the parts are maintained, thereby maintaining the properties of the system throughout the change (*SC*, 50/47). In this way, a form is a "total process" that serves as the organization of the reciprocal action of the parts upon one another, a way of explaining the mutual determination of the parts by the parts in such a way that there is a specific distribution of energy within the system given a determinate stimulus upon it. Form is not a *result* of causal, physical processes and does not exist as a being. Rather, "the notion of form only expresses the descriptive properties of certain natural ensembles" (54/51). The perception of a natural ensemble, a group of processes that coheres into a single perceived process, occurs only through the mediation of a form.

If a form is not an existing thing, then the question arises as to its ontological status. At this point in Merleau-Ponty's philosophical development, he responds as follows: "Thus form is not a physical reality, but an object of perception; without it physical science would have no meaning, moreover, since it is constructed with respect to it [form] and in order to coordinate it" (*SC*, 155/143). Perception, of course, should not be taken as merely something immanent to a consciousness, for that would be to revert back to an intellectualist position. Perception, rather, always extends outside the body into the world of things. Ted Toadvine introduces a distinction between intellectual and perceptual consciousness in order to clarify this point. Intellectual consciousness "is oriented toward the one universe shared by all rational subjects," but it is "derivative from a more fundamental perceptual consciousness, which remains immersed in its environment and lacks explicit self-awareness."[10] Anonymous perceptual consciousness, the openness to the world that makes contact with beings, perceives an "internal unity" to the relations between itself and other beings. In perception, though, he is going to want to retain the limited truth of both naturalism and the critical tradition. Insofar as perception is *of* and *in* the surrounding world of things, it depends upon those things to show themselves through their relations with the perceiver and other beings. As something present within the perception of a living being, however, the internal unity of form will remain relative to the perspective of the perceiving consciousness. Form is *real*, but is real *as perceived*. Because the world is ordered in perception, and perception, as perspectival, creates meanings, order arises from the way in which certain processes display a directedness toward other processes. The ordered, natural ensemble, a form, then, exists only in the relationship between a world and a consciousness. In *The Structure of Behavior*, Merleau-Ponty's point is to indicate that everything that appears to consciousness appears to it within a structured system of relations that provides an order to the perceptual field in the form of natural ensembles. These ensembles in turn have a meaning precisely because of the organization of the perceptual apparatus, a physical given, and the relations between beings in the field. There is, therefore, a very narrow interrelation between form, consciousness, and the physical world such that order is attributed to that world on the basis of perceived wholes. In many ways, this relation between the three terms is what Merleau-Ponty will later call the "primacy of perception": all of our philosophy and science has its basis in a specific understanding of perception.

The relevance of behavior to our investigation is that, on its basis, we can begin to see how nature can be conceived of as a regular, ordered, and meaningful whole while refraining from the ascription of a mechanistic or teleological character to that whole. Behavior functions within a given system; it occurs through the distribution of forces within a given structure. Behavior occurs, for Merleau-Ponty, on the basis of perception because, as a structure, it relies upon a perceptual field. The perceiving being exists within a field of perceived and distributed values, within which the perceiver is but one element of the larger perceived system. Merleau-Ponty calls this meaningful, perceptual field the body's "situation." From its situation at any time, other forces within the field are constantly acting upon the body, calling for a response from it. There are, however, myriad ways in which a body can respond to a given stimulus rather than a single, fixed, reflexive response. Behavior is the spectrum of possible responses to the influences that affect a body, but the crucial point to note is that, while the spectrum is fixed to a certain degree due to the organization and comportment of the body, the response is essentially open. Outside the artificially conditioned environment of the laboratory, most beings exhibit a much broader spectrum of behaviors and potentialities for learning than is usually attributed to them. To demonstrate this point, Merleau-Ponty refers to the phenomenon of *learning* (*apprentissage*).[11] In this case, learning does not refer to the mere incorporation of a new stimulus-response function, which is how many comparative psychologists describe the successes of their subjects. Rather, in terms of this theory of behavior, learning is not *merely* repeating the same gesture with respect to the same situation; learning "is a matter of a new aptitude for resolving a series of problems of the same form" (*SC*, 106/96). The world presents obstacles to the body's activity that fall within a given number of forms, each of which has a specific set of behavioral responses that are applicable to it. A behavior is a habitual way of confronting a problem of a determinate form learned through continual confrontation of problems of that perceived form over time. In order to respond to the problem, the situation must be perceived in a way that allows for a response that addresses the problem. The acquisition of a new behavior, then, occurs within a given perceptual field that is rich in already acquired meanings. A behavior is learned when new forms, that is, new structures within the perceptual field, are recognized and in so doing new meanings are discovered within the perceptual field. Thus, behavior is never only a physical event, but rather "becomes the projection outside the organism of a *possibility* which is internal to it" (136/125). Quite simply, even if learning were mere rote repetition (i.e., stimulus-response), behavior would still require that the situation be *appreciated* as one in which the behavior is applicable. This appreciation is possible, in his view, only through the notion of form. Thus, a behavior is also a determinate way of distributing values, in this case perceptual values, within a given field of experience so as to achieve certain goals.

One will note that behavior has been described primarily in reference to perceiving and thus living beings. A lingering question, then, is how the notion of behavior helps us to formulate a nonmodern conception of nature. Though Merleau-Ponty himself does not follow this path, the conjecture that I would like to develop for the remainder of this chapter is that his notion of behavior can be applied to natural phenomena precisely because the formulation of the concept of behavior relies upon a relational ontology. If behavior can be described as the ability to discern distinct forms within a field that structure the field and give it meaning, then there is a sense in which inorganic beings accomplish similar feats, albeit unconsciously. Merleau-Ponty himself opens up the possibility for this conception of nature, but not here in his early philosophy. In order to see how and why one might desire such a conception, we must conclude this section by looking at the vision of nature from which Merleau-Ponty eventually diverges for the reason that the basic structures of his later thinking are already present within it.

For Merleau-Ponty, that to which one traditionally refers when one speaks of nature is one order among three. Nature, or the physical order, is the foundational order upon which the other two supervene, but at the same time it is not sufficient to explain the existence of behavior and mind, which belong to the vital and human orders, respectively. Insofar as each is an order, each is characterized by a distinct form, which means that certain relations that are efficacious within a given whole might not be efficacious in others. In other words, there are events within the physical order that in principle have no influence upon the human order (e.g., human vision is not sensitive to infrared light or, for the most part, heat), and vice versa (e.g., one can imagine Pegasus without this having any effect in one's environs). In this understanding, each order is determined in terms of a dominant characteristic for that order: the physical order is primarily quantitative, the vital order is valuing, and the human is signifying (*SC*, 141/131). At the same time, however, because the orders are mixed together in the apprehension of a living being, the dominant characteristic of each is still a property of each of the other orders. Naturalism's attempts to reduce all phenomena to a physical substrate are misguided, not because there are some phenomena that do not have their basis in the physical world, but because the reduction of value to a physical process misses precisely that which makes something of value: a vital relationship to a world. The same is true for mental phenomena: there are physical events that subtend mental events, but to identify the signification of a mental event with the physical event destroys the signification, that is, fails to recognize the cognitive function that the meaningful relation as meaningful has within the perceived world.

But if the world is constituted through the three perceived orders and the three are mixed together and isolable only in thought, by which I mean that one can segregate and discuss them separately only by means of analysis, then the question remains of what difference sets them apart from one another such that they are not in fact reducible. The orders are related, but each of them, in order to be the order that it is, needs to be somehow distinct from the other orders. Merleau-Ponty puts the thought this way: "Matter, life, and mind must participate unequally in the nature of form; they must represent different levels of integration and, finally, must constitute a hierarchy in which individuality is progressively achieved" (*SC*, 143/133). Merleau-Ponty is going to separate the three orders based on the degree to which they reflect the integrated, ecological ideal of a whole of nature. The physical order is the least integrated insofar as it is the order most in need of supplementation in order to account for observed phenomena, and it is also the least endowed with individuality given that this order is primarily marked by quantitative relations. For this reason, the physical order is the base of the hierarchy of orders.

It is important not to interpret the assertion that the physical order is the first in a hierarchy of orders in terms of a monistic metaphysics since Merleau-Ponty's ontology is not based on substance. To say that the physical order underlies the other two is not to assert a materialist reductionism for two important reasons: (1) reductionism is predicated upon the existence of a substance to which the higher-order phenomena can be reduced; and (2) the higher-level orders, once they emerge, serve as an independent reality of meaning for the animal or human being. Matter, as Merleau-Ponty speaks of it, is *another form*. So not everything that exists can be understood entirely in terms of the relations between matter, even as everything that exists participates in physical forms. This idea can be difficult given our propensity to understand the world in terms of substance, so Merleau-Ponty explains the idea in greater detail: "In a philosophy which would genuinely renounce the notion of substance, there could be only one universe which would be the universe of form: between the different sorts of forms invested with equal rights, between physical relations and the relations implied in the description of behavior, there could be no question of supposing any relation of derivation or of causality, nor then of demanding physical models which serve to bring into being physiological or mental forms" (*SC*, 144/133–34). What we see here is that one cannot expect to discover a causal relationship between the physical order and the higher orders, but if we recall that form is *a perceived internal unity of relations*, we can also note another point here: "that matter, life and mind could not be defined as three orders of reality or three orders of beings, but as three planes of signification or three forms of unity" (217/201), a point he had already introduced earlier in the text (147/137). So, to a large degree, Merleau-Ponty is still operating within the critical tradition, though he will also distance himself from certain aspects and interpretations of it by integrating naturalistic elements as well.[12] The philosophical meaning of "form" is not in "the idea of signification but that of *structure*, the joining of an idea and an existence" (223/206); so, in calling the three orders different planes or levels of signification, Merleau-Ponty is indicating how there will always remain a direct connection between meaning and being: "The signification which I find in the sensible whole was already adherent in it. . . . The signification is embodied" (228/211). The key, then, to understanding the surrounding world as form is to embrace the situatedness of perceptual consciousness among the things in the world, "to show a mind which *comes into the world*" (225/209). Forms, as expressions of the perceived world, maintain a relation to that world and thus an indirect reference to that which inspired such a form. Again, intellectual consciousness does not constitute beings because the forms are perceived relations within the world; if they are to exist, the synthetic a prioris of transcendental philosophy will be located in the body, not an intellectualized consciousness.

What remains for us to discuss here is how forms present themselves in the physical order. The function of physical systems is most accurately described through the physical sciences. Since scientific observation is a specific mode of conscious interrogation of the world, scientific laws must be acknowledged as being specific, more rigid forms that are expressed within the perception of the world by a scientific observer (*SC*, 112/102). The inherence of the physical law in the consciousness of the scientist does not mean that the law is itself "relative" or "constructed" by the scientist, however, only that the law is expressible as a law through the mediation of symbols and meanings that themselves have human, cultural origins. Rather than having direct access to "reality," then, science offers a view of the world within which the perceived natural ensembles are laid bare in their quantitative distribution of forces, but only through the mediation of symbolic forms. More simply put, science can describe only what it sees, and since all perspectives, by definition, are limited and are encompassed by a horizon, science cannot completely disclose to consciousness the world as a whole, but can describe only the forms through which we interact with nature (235–36/219). As an example, one can look at the history of two distinct physical, scientific concepts, gravity and light, in order to see how variations in their formalizations can lead to the construction of distinct pictures of what the world is like. Einstein's theories would not have been as momentous as they were had they not challenged the very heart of classical physics through a *reformalization* of gravity, and, on the basis of his reformalization, the picture of the universe that physics advances altered dramatically, introducing such unheard-of phenomena as black holes, dark matter, and the previously discussed strings of string theory. Also, by means of subsequent developments within physics, a predominantly probabilistic model came to replace the mechanist, classical model of nature. The world did not change, nor did the observed phenomena: what changed were the equations, the different formulae used to express the observed relations, and thereby the very consistency of the network of relations that is said to constitute the physical world. Thus, physics constructed a new *picture* of the world in order to account for a change in the constitution of the system. To this extent, Merleau-Ponty's account of the knowledge of the sciences is consonant with that of Heidegger, though he will find a different meaning in the expressive character of the sciences.

The vision that Merleau-Ponty advances of the physical order follows quite closely how the science of ecology envisions itself today: forms are present in the physical world to the extent that physical systems, which he defines as nothing but ensembles of forces in a state either of equilibrium or of constant change (*SC*, 147/137), possess an "internal unity" that "resists deformation," and exhibits a "circular causality" among its parts (148/137). To the extent that a given system possesses this internal unity, it can be identified as an individual physical form. Thus, within the physical world, there are even varying degrees to which a form can be called integrated: an individual photosynthetic effect is distinct from, and yet intimately tied to, the manner in which the atmosphere maintains its chemical composition. What is to be noted is that forms are always relative to the level at which one is inquiring: if what is of interest is the photosynthetic effect, the individual form will be within the leaf itself; if what is of interest is how the chemical composition of the atmosphere is maintained, the individual form will be the totality of relationships that contribute to the production of the atmosphere.[13] These systems, however, are historical insofar as they are alterable once the relations that obtain between the individual beings within the system are pushed beyond certain thresholds: differences in degree can inspire a change of kind via a redistribution of the forces into a qualitatively different order (148/137). To note this, however, is to introduce the idea of discontinuity into the physical sciences, to assert that historical relations have an influence upon the integrity of physical systems. Of course, this idea is not so foreign today given the current ecological conceptions of "resilience" and a greater understanding developed over the course of the twentieth century of the fact that ecological systems do not tend toward an end or climax state. If one accepts that physical laws are the formalized expression of the properties observed in the surrounding world, then these laws are indicative of specific forms of certain relatively stable ensembles, some more stable than others. The formalization of gravitational force is accurate only given a whole series of factors that are taken ceteris paribus and, under usual conditions, do not need to be questioned (e.g., the mass of the sun and earth, the velocity of the earth's rotation, etc.). But *if* the sun or earth were to somehow lose half its mass, it is an uncontestable fact that this would alter dramatically every one of the physical systems on earth and all the formalizations of the relations within them. Physical laws, then, are not immutable and given in advance for the scientist to discover, but expressions of the stability of a given ensemble. A physical form, then, is the isolation of one such natural ensemble that exhibits a high degree of stability. Merleau-Ponty calls these "partial totalities" (149/139), and, properly speaking, the notion of a partial totality is what replaces the outdated view of nature as "*omnitudo realitas*." Rather than conceiving of nature as an actual *thing*, it must be understood as a historical reality. We can grasp its ontological status in the way Merleau-Ponty describes that of ancient Egypt: "It is an idea, a signification, common to an ensemble of molecular facts, which is expressed by all the facts and which is not contained completely in any of them" (155/143). As signification, however, the idea of nature does not belong to the physical order. "Nature" originates in the human order as a way to manage and organize the perceived partial totalities. Nevertheless, we should not hear this designation of nature as a signification or idea to mean that nature no longer has any reference to the physical world because the idea of nature organizes the ensemble of physical partial totalities.

Nature exists, then, because a conscious being perceives partial totalities within the surrounding world and perceives an organization amid the relations between them, keeping in mind, of course, that these partial totalities are perceived because of concrete relations in the world. Nature involves, therefore, a double *dialectical* relation: one dialectic takes place between physical forms and physical laws, the other between perceived physical forms and the consciousness that discovers a signification therein. The first dialectic has already been discussed insofar as there is a constant reciprocal determination of exchange involved in the continual formalization and reformalization of laws given shifts in the conditions that give rise to those formalizations. In short, there is not, and in principle can never be, an absolute expression of nature's laws because elements of the system are constantly undergoing change. The second dialectic occurs at the level of signification, not at the level of the physical process: the unity of the partial totality is *perceived*. To see the consequence of this assertion, consider the question of the limits (or even existence) of an ecosystem. If we equate an ecosystem with the notion of a partial totality, then it will exhibit a certain set of reciprocal relations that establish a relative equilibrium. But the equilibrium of the internal relations is perceived only upon the background of outside forces that act upon it, that is, other partial totalities. The ecosystem is an ecosystem because it maintains its internal relations in the face of external influences, up to a certain threshold. These smaller ecosystems can in turn be organized into a single partial totality that maintains a relative equilibrium with respect to outside forces. For Merleau-Ponty, the ensembles are meaningful wholes, that is, an ensemble is isolated only through the mediation of signifying structures. For example, whether a scientist identifies a salt marsh as an ecosystem, or incorporates it into the larger system that encompasses the nearby bay or estuary that provides the salt marsh system with nutrients, depends largely upon what kind of data is being sought. The point here being that how the physical systems are divided is at least partially dependent upon the world of signification of the perceiver. However, it should also be reemphasized that the systems themselves must be there to be perceived. There is, then, a dialectical exchange between perceived beings and the signifying world within which they appear, while the lower levels do not integrate the activities of the higher.

So what is left of naturalism in this view? Although it strips the physical world of normativity—norms are derived from values, which are of the vital order—structure joins together and makes indiscernible idea and existence, but only within a given field (*SC*, 223/206–7). At the same time, however, Merleau-Ponty explicitly endorses the Hegelian definition of nature as the exterior of the Concept (175/162, 227/210), which accords with his own assertion that nature has a meaning only for a human consciousness; form is real, but is so only within consciousness. Insofar as humans are humans, though, nature does indeed exist as the limit idea with respect to the organization of physical systems. We must not, however, take this idea of nature to be radically anthropocentric because the signification "nature" is grounded in the apprehension of the order within the world outside consciousness. In other words, even if nature is a signification and the forms we find within it are real only within consciousness, all forms are derived from an "original text," the physical systems observed in the world, and, as such, meaning is derived only from the world (228/211). The ground of signification is concurrently in nature and in consciousness, which is to acknowledge the philosophical import of the dialectic between perceiver and perceived. Significations are no longer "in the mind" because the order-based ontology does not recognize a distinction in substance between body and mind. Signification is always already in the world, which is why Toadvine can say, "Nature at its most fundamental level is meaningful and experiential; its structures manifest a unity and coherence that characterizes perceptual wholes."[14]

With the integration of signification and existence within the perceived world, Merleau-Ponty draws his only explicit account of what nature is to a close. This fact poses a distinct problem for us insofar as his metaphysical views put forth here differ meaningfully from those he advances later in life. Accordingly, these views of nature cannot be called his definitive views. Yet, given the lack of another text that might supply us with anything more than clues as to what his conception of nature might look like, what remains of this chapter will be largely reconstructive and speculative in content. I mention this not to dismiss Merleau-Ponty's own intentions and convictions with respect to the importance of nature as the "soil . . . that carries us" (*N*, 20/4), but rather simply to make clear that these ideas are drawn from Merleau-Ponty's thought and are not necessarily, strictly speaking, what he would have developed and set forth if not for his untimely death.

Disclaimers aside, the early view opens many questions for thinking that had not previously been viewed as questions one would ask with respect to nature. If knowledge of nature arises through the mediation of forms, a relevant question to ask is how the natural world expresses itself such that the form arises from the dialectic between external world and consciousness. In other words, how does a perceiving being apprehend the signification latent in the world's physical processes? If his view is not merely a traditional return to Kantian critical philosophy, which Merleau-Ponty continually emphasizes it is not, we need to examine more closely the structure of expression and how it removes his philosophy from the immanence to consciousness both Heidegger and Latour criticize.

II. READING THE PROSE OF THE WORLD

Merleau-Ponty addresses the above issues in the unpublished manuscript *The Prose of the World* in a way that demonstrates the direction of transition in his views.[15] Though one cannot be certain as to why Merleau-Ponty abandoned the manuscript, when one compares the text with writings published during his lifetime, it becomes evident that he did not in fact abandon the ideas contained therein. Evidence for this lies in the publication of "Indirect Language and the Voices of Silence," which was adapted from the third chapter of *The Prose of the World*, titled simply "Indirect Language," and the presence of passages from the manuscript that appear in various other publications such as "Eye and Mind."[16] *The Prose of the World* remains instructive on its own insofar as it addresses two key concerns that will segue from the earlier view to the later: a refinement of how the structure of expression works, with explicit reference to the interplay of nature and history, and an elaboration of the part intersubjectivity plays within both nature and history.

In *The Prose of the World*, Merleau-Ponty approaches the questions from *The Structure of Behavior* through the lens of language and expression. The reason for addressing the question of nature in this way derives from the paradoxical conclusions of the earlier work: if nature is a *signification* of the human order that helps human consciousness to organize the perception of physical processes and systems, then the essence of nature is to be sought in the source of the human signification, the phenomenon of expression. Contrarily, he makes the assertion that the meaning of the signification is derived from the processes themselves, which is to assert a truth to naturalism as discussed above. There is an obvious tension between these two views insofar as they unify what have been traditionally considered both subjective and objective elements. By asking the question of how the "objective" expresses the meaning of the "subjective" phenomenon, Merleau-Ponty demonstrates that these two positions, that of the subject and the object, are consistently undermined by a relational ontology because objects are constituted by a subject, while at the same time the subject is constituted by the relations it enters into with beings outside itself. The circularity of these relations where neither one is able to claim precedence and priority over the other leads him to seek the point at which the two dualistic poles are still yet undifferentiated in order to see how and why they separate. Here we can see him, like Latour, attempting to undermine the pure subject-object distinction as a bad way of articulating the relationship between humans and nonhumans. The structure of expression is the means that Merleau-Ponty will utilize to undermine the opposition of subject and object, nature and history, and, like Heidegger, attempt to open another history within which nature will be conceived otherwise than in the tradition of Western philosophy and science.

We will begin our examination of expression by returning to the naturalism advanced at the conclusion of *The Structure of Behavior* in order to determine the extent to which nature and consciousness respectively provide meaning to the signification of nature. Merleau-Ponty begins, as is frequently the case, with a paradox. If language allows a direct correspondence between a statement and a referent, then it can do so on one of two bases. The first possibility gives the power of correspondence to the things. This view holds that there is a basis for language among the things themselves, a "language spoken in things," and so language, as a tool for human use, can be made adequate to this nonhuman language of nature (*PM*, 11–12/6–7). What he has in mind here is the mathematical ideal of reducing the movement of nature to a human-generated system of signs (numbers and variables) that accurately describe observed natural processes. The alternative possibility is to give the power of correspondence to language itself. This view holds that language is a completed system, that is, it contains all its possibilities within itself in advance in the form of given signs, permutations of pregiven expressions and signs, and possible combinations between signs and signifiers (12/7). In this case, because a language already contains within itself the possibility of expressing all possible relations, achieving correspondence is simply a matter of finding the correct meaningful expression for a given perceived phenomenon. The former is a philosophy of the pure sign, holding that a sign represents exactly, without surplus or loss, the phenomena that it indicates, while the latter is a philosophy of pure signification, reducing the concreteness of that which is thought about and spoken of to its manner of representation in language.

The philosophies of the pure sign and the pure signifier both result in the same outcome: the vacuity of communication. In the case of the pure sign, communication becomes unnecessary because everything that would be expressed is already contained within the original perception. Language becomes superfluous because the perception already says everything that is to be said of the phenomenon in a more pure and precise way than human expression ever could. In the alternative case of pure signification, communication ceases to address an outside world: signs do not refer to objects in the world, but rather to significations in the minds of other language-users (*PM*, 13/7). The world, in this latter case, becomes superfluous because what is essential to communication is the transmission of true significations from one mind to the other, not the description of phenomena. So, if these philosophies are seeking to explain how language and the world are capable of correspondence, they both fail by sacrificing one of the terms of the relation to the primacy of the other. Merleau-Ponty draws the conclusion from this analysis that there can be no direct relation of the world and language, but rather that there must always be a third term through which they will be related.

This last conclusion leads him to assimilate the basic principles of Saussure's structuralist linguistics, within which all signs are diacritical signs, that is, they do not themselves directly express or represent anything but take up a signifying value from their place within a system of signs. If this is the case, expression is never direct; language never directly refers to an external object, but rather constructs a system of "mutual differentiation" between the signs such that, with reference to a third term mediating the two communicants, a meaning emerges from the signs (LI, 65/242). Communication becomes possible between two speakers because the system of mutual differentiation is common to both and *mediates* their communication, "propels us toward the things [language] signifies" (*PM*, 16/10). The two communicants are not each confined within a private world of signification, but carried outside themselves into the world because it is only through the mediation of things that it is possible for them to speak to one another. The mediation of the perceived world, that is, the mutual relatedness to the same phenomena, is what makes communication possible. Though Merleau-Ponty does not use the term, we could call this mediation the "mediation of things." The mediation of things aims at replacing the philosophy of the pure sign by removing the sign from nature and placing it within the human realm of signification.

The mediation of things, however, is not yet sufficient to describe the phenomenon of expression. After all, this type of mediation seems to be nothing more than the reference view of language that Merleau-Ponty eschews: if language is the creation of a ciphered version of an original text, then all language is a translation from the language of things into the language of humanity (LI, 69–70/244–45). One can already see the partial rejection of some of his themes in *The Structure of Behavior*: Forms can no longer come from the direct experience of an "original text" of the world because such would result in a view of language as translation rather than as a means of differentiation and articulation of the world. In the translation view, meaning is derived from the speakers' common referent, and the signifier becomes an arbitrary symbol meant to adequately express the thing itself. It does not matter if I use the words "the place I sleep" or the word "home" to describe my residence because both phrases pick out the same object in the world. Of course, in everyday communication this choice means a great deal: calling one's residence "the place I sleep" connotes a very different and less welcoming relationship than does simply calling it "home." The difference between these two senses is precisely what is communicated through the choice of phrases, since in my choice of words I am expressing a particular kind of relation to a thing, not indicating the thing itself. To communicate the relation, a second kind of mediation is necessary, that of institutions. It is through the notion of institution that Merleau-Ponty will try to reintegrate history into nature such that the two become indissociable.

Once introduced, institution becomes an increasingly central part of Merleau-Ponty's philosophy, though it is already present in his thought in the distinction between speaking speech and spoken speech in *Phenomenology of Perception*, which appears again in *The Prose of the World* (*PM*, 17/10). Spoken language is "an already shared common terrain of acquired and available significations" (18/11), a "stock of accepted relations between signs and familiar significations" (20/13). As such, spoken language provides the forms through which perceivers interact with their world because the signs do not refer to things, but to relations. Speaking language, on the other hand, is an initiation into another world of relations, the attempt to take up the spoken language and to *rearrange* the relations in order to make possible novel significations. To the extent that speakers construct a system of relations immanent to their speech, the system of relations that constitutes the listener's world, here in the Heideggerian sense, is challenged by a related and yet foreign set of relations. The transformation in the system of relations is quite literally a change in perceptual form, allowing the perception of new structures and organizations that had hitherto remained concealed. This transformation comes, of course, at the cost of covering over certain other relations that had previously existed. Speaking speech, however, advances a variation of the common relations, alters the cultural terrain, while one is speaking. Eventually, however, speaking speech recedes into the background of that cultural world and becomes a part of the given set of relations into which the speaker is initiated, which is to say that speaking language always becomes spoken language. This effacement of language by language, the receding of the *formative* aspect of language into latency, is an unavoidable consequence of the operation of language, and without it there would be no cultural inheritance of language: we hear what is said, the signification, not the saying of the said (15/9). The effacement of language, however, is precisely what makes possible the traditional philosophies of language of the pure sign or pure signifier and what makes so foreign the view of language as a historical means of expression, an "endless proliferation of aberrant forms" within which "past, present, and future are mixed" and "no rigorous break is possible" (56/39). For language to be a proliferation of aberrant forms, however, there must be the spoken language from which speaking language can diverge: creation and novelty requires a cultural milieu in order to create its own milieu. This structure of divergence, which Merleau-Ponty will later call *l'écart*, is fundamentally historical in the same way that authentic history is, for Heidegger, the history of the transitions in the constitution of world. Speaking speech, then, *inaugurates* a new world (60/42); it opens up a horizon of meaning that did not previously exist, a dimension within which humanity can then dwell. "Institution" means the inauguration of a dimension within being, dimensions of meaning that mediate our relations to things and others. Institution, then, provides a second level of mediation through which particular forms are established within perception and things are seen in terms of these historical forms. If, as Latour claims, the nonmodern world is marked by its concern for mediation over purity, focusing further on the nature of mediation operant here will be helpful in developing a nonmodern understanding of nature.

In the concrete act of communication, then, there are two distinct types of mediation at play: the mediation of things and the mediation of institutions.

What remains is to see how the two are actually implicated within each other and separable only in reflective analysis. The two types of mediation are actually a single dialectic of experience, a movement back and forth between two poles of interpretation that require each other and reinforce each other's position. While expressing the point this way makes it sound as if Merleau-Ponty falls prey to Latour's criticism of phenomenology as producing an irresolvable tension between two dualistic poles, the situation is actually quite the opposite. The mediation of things and the mediation of institutions always come together and for that reason form a unity. The union of these two mediations can be seen most clearly in the discussion of culture and its relation to the perceived in "Indirect Language and the Voices of Silence." The key point is that while perception is a natural function, it is always performed within a culture (LI, 78/250). So Merleau-Ponty can also be said to believe in something like nature-cultures. The reason for this lies in the horizonal nature of perception. As discussed earlier, perception always occurs within a given field, a field that is articulated in specific ways depending upon the forms that are perceived within the field. The perceptual field, then, consists of things perceived through forms, which in turn make those things appear as meaningful. Now, if we incorporate into this model the structure of institution, the picture becomes less that of a constituting consciousness projecting forms into a world that corrects or adjusts those projections and becomes more that of Heidegger's thrown thrower: the perceiver never has control over how the world is articulated because the perceiver always already finds itself within an instituted, significant world. We already described this structure in terms of spoken language wherein a speaker is born into a context of relations to things. Here, however, the structure is more general insofar as the spoken language–speaking language dichotomy can be grafted on to any situation to which a living being responds. The concept of behavior comes to mind: a situation is presented within which there are various possibilities for response projected outward from the living being. A being may pursue any one of these reified pathways of response, or the being may behave otherwise in a novel manner that opens up distinct pathways for future possibilities and the potential to recognize new forms. Behavior takes place with respect to institutions, either within them, according to their internal logic of articulation, or alongside them, transforming them by means of the inauguration of another dimension of behavior. The same is true of perception as is true of behavior. If perception operates according to given perceptual forms, then we must also admit that the forms are received in advance through specific institutions. But note that the institutions through which perception is possible are both physical, since the constitution of the body is the "primordial institution" that makes all symbolic articulation possible (*PM*, 63n/45n), and cultural, since the field received through the senses is articulated in meaningful ways.

If culture informs perception, then cultural practices can change the way we see. On the model of institution, this exchange occurs through the inauguration of new behavioral dimensions. Merleau-Ponty's privileged and preferred method for demonstrating this point is through painting, though the same could be said of all forms of art. What a painting accomplishes is to teach a new way of seeing the world, a new way of articulating encountered beings, and thus a new way to think about them (LI, 90/256). A painting achieves this through the creation of a coherent system of equivalences within itself that is capable of decentering the already established system of equivalences within spoken language and thereby itself engaging in speaking language (91/257). This inauguration, or, as he alternatively says, instauration, of a new system of equivalences is what allows Merleau-Ponty to say that paintings are the "voices of silence," insofar as a painting engages in linguistic expression without even uttering a word. In this way, we are invited to look at all means of expression as linguistic insofar as they possess the speaking language–spoken language structure. As a result, all bodily gestures, as instituted or inaugurated responses to a perceived situation within which the body finds itself, are considered to be linguistic gestures. The expansion of the linguistic universe will be a very important aspect of the reformed view of nature for the reason that it shows the extent to which nature, as a signification supplied to physical processes from the human, signifying order, is intertwined and inseparable from culture.[17]

On the basis of the mediations that intertwine nature and culture, Merleau-Ponty is able to overcome the immanence of the traditional phenomenological consciousness, but not without raising new difficulties. Because perception happens on the basis of the primordial institution of the body, which is in turn instituted into the cultural-historical order that supplies forms to perception, the "I think" of traditional critical, transcendental philosophy must be rejected because the "I think" can exist only through these prior institutions. The pure "I think" that underlies all judgments and serves as evidence of humanity's supersensible nature within the Kantian system is, in fact, aware of itself only through the mediation of a body that finds itself in a world and over whose development and genesis the "I think" exhibits no control (*PM*, 27/18). Thus, the "I" of the "I think" "carries in itself the germ of a depersonalization" because it requires the existence of a prior stratum that is anonymous and outside volitional control (29/19). The instituted essence of the self calls for the abandonment or complete reformulation of the concept of autonomy in order to account for the myriad ways in which one's actions are not one's own because they are, in fact, behaviors operating within instituted systems.

In addition to being a historical structure, institution is an intrinsically intersubjective phenomenon. Communication, the phenomenon that led us to these reflections, requires that there be something or someone who is addressed. In *The Prose of the World*, Merleau-Ponty attempts to break out of solipsism in the same way that Husserl does in the fifth chapter of *Cartesian Meditations*: because other human bodies appear similar to my own and respond to the world in a similar fashion, they, too, must be conscious in a way similar to my own. As with Husserlian empathy, there is no direct access to the consciousness of the other person because empathy occurs only through the mediation of a series of gestures and exchanges between the two. But remember, communication carries the speaker outside itself insofar as language addresses a world external to consciousness and takes place through a system of meaning that does not originate within consciousness; gestures address beings in the world. For this reason, Merleau-Ponty hypothesizes a "universality of feeling" (*PM*, 191/137), upon which the experience of a *body* can be generalized, not a separately subsisting consciousness. The universality of feeling introduced here is in many ways a precursor to the later concept of flesh insofar as it introduces *affect* as the medium through which different bodies relate. For now, what is relevant is the binding together of the two bodies in a medium. If two communicants are to respond similarly to a situation, the situation must be common to both of them, that is, the perceptual field of another human being must be articulated similarly enough to my own that a gesture made toward a being in the world is comprehensible to me. The identification of the other as a subject, then, does not occur on the basis of any intrinsic properties or characteristics that they might possess, but rather on the basis of their behavior toward a perceived world. But in order to behave, one must perceive. Hence, if there are other conscious bodies out in the world, there must also be the universality of feeling that is articulated by both of us: each of us has a perceptual field that is a variation, based upon our position and cultural situation, of the other's. But the other maintains their difference because the other's perceptual field is precisely their field. I can never see what the other sees because our fields can never coincide; we will always diverge by temporal, spatial, or cultural differences that are all lodged in the body. Intersubjectivity, then, is a complex relation with multiple relata: there is my body, the other body, the natural, sensible world that we both address and to which we both respond, and the cultural institutions that mediate our relationship to both each other and the world.

What the phenomenon of communication makes plain is the *commonality* of the world as the multiplication of perceptual fields within a universality of feeling. When Merleau-Ponty speaks of the common, he is not merely deploying Plato's conception of the forms as what is common (*koinon*) to all beings of a certain type. He is invoking Plato's view with the intention of reformulating it. Commonality is not found between distinct properties of a being because those properties are relative to a perceived field. Commonality lies in the situation within which beings are the beings that they are and are perceived as possessing the properties that they posses. More simply, commonality no longer refers to well-defined beings sharing specific properties, but to the relations that provide a being with its ipseity. Even this is slightly misleading, however, since aseity is more properly attributed to beings than ipseity:[18] a being never is, but is always becoming what it is through interaction. Merleau-Ponty explains this most clearly in the "Sensing"[19] chapter of *Phenomenology of Perception*. The context of the discussion of a being's ipseity is Merleau-Ponty's argument against the existence of a mental, epistemological subject in favor of perceptual intentionality based on the primordial, latent institution of the body. The thrust of the argument is that the senses are not unified in a mind, but rather are all experienced together outside the mind in the being that is encountered: "The senses communicate with one another in opening onto the structure of the thing" (*PhP*, 265/229). The thing, then, is a "significative nucleus" (266/230) that unifies the diverse perceptual intentions of the body, but this nucleus is not invariant insofar as perceptual intentions are always situated and from a particular perspective. If the thing perceived, however, is taken to be persistent, if its reality consists in its being constantly present, then the essence of the thing will be found among its fixed, perceived properties. The operant assumption here is that the thing contains its nature within itself, that is, its essence lies in its ipseity. If, on the contrary, one adopts a relational ontology that does not begin with the idea of substance, then the changes in properties that correspond to the changes in perspective make this assumption of ipseity questionable given that the perceived thing alters its properties under changing conditions, both those in the perceiver (e.g., fatigue, jaundiced vision, etc.) and in the world (e.g., changing lighting, movement within the field, etc.). Such is the basis of John Locke's attempts to ground his metaphysics on so-called "primary qualities" and George Berkeley's counterassertion that primary qualities are known only through subjective, secondary qualities. Merleau-Ponty sides with Berkeley in that dispute, but draws a distinct conclusion from this fact. Rather than displacing the ipseity of the thing into a different substance, the mind of God, Merleau-Ponty asserts that perception is a process, and therefore "ipseity is never *attained*" (*PhP*, 269/233). If this is so, the reason is that the thing is not a substance but a nucleus of relations; it has its nature outside itself in its relations. So, rather than assume the ipseity of the thing as completed and therefore require the inference of an unchanging substance to underlie all the perceived changes, which is Kant's way of resolving the tension in the first "Analogy of Experience" in *The Critique of Pure Reason*, the reality of the thing is situated in "that which hides it away [*dérober*] from our possession" (270/233). The word *dérober* is quite important here because it indicates the affinity that Merleau-Ponty has early on for Husserlian phenomenology, within which beings in the perceptual field are given as adumbrations (*Abshattungen*), and how this idea will undergo a change in sense once he recognizes how Heidegger will generalize this perceptual, phenomenological observation into an ontological structure. At this point, however, Merleau-Ponty is referring to the fact that a phenomenon is constituted by the totality of relations in which it engages, meaning that a thing *is* its relations. Once again we see the phenomenological connection to Latour: beings are events. As such, the thing possesses an aseity, which Merleau-Ponty defines as "its irrecusable presence and the perpetual absence into which it [the presence] is withdrawn" (270/233). In other words, because the thing appears only from a perspective and there are, in principle, an infinite number of perspectives from which the thing may be observed, the thing's ipseity is always in becoming or is a process, which requires that certain of its aspects become absent when other of its aspects become present. Put into the language of Platonism, the thing no longer receives its form from the intelligible world in which it is also said to participate, but rather the thing receives its form by participating in a sensible field that consists of other sensible beings.

Taking a broader view for a moment, Merleau-Ponty's philosophy can be described as a reworking of Platonic *methexis* (participation). Conceiving of Merleau-Ponty's philosophy in this way is interesting, in terms of the current argument, for two reasons: (1) it highlights the connection between this form of phenomenology and Latour's thinking about mediation; and (2) focusing on this aspect of his thought will help us to understand that, like being, mediation has a history, which is what Latour was accused of forgetting. While it is not possible to do justice to the two millennia of discussion surrounding Plato's theory of forms, the doctrine of participation is Plato's solution to the problem of reconciling the relation between the sensory world of becoming and the eidetic world of being.[20] Sensible beings exist in an intermediate realm between being and nonbeing and so maintain their being through a relationship with what is really real, the forms. This relation between form and sensible being is participation, and the participation in various forms is what makes a being what it is throughout the changes it endures within the sensible world. The essence of a being, its reality, lies outside its sensible existence in the eidetic world with participation serving as the bond between them. The nature of this relation is somewhat mysterious, much like the presumed relation between mind and body in Descartes's philosophy.[21] What can be safely said, though, is that the visible world exists in virtue of its participation in an invisible world.

In reformulating participation in terms of a relational ontology rather than one based in substance, Merleau-Ponty will understand the relation between being and becoming, the visible and the invisible, very differently. As for most phenomenologists, there is but one world, common to us all, within which all beings participate. Their participation in the world happens at multiple levels: they participate both in their mutual becoming and in the formation of the world itself. The first type of participation is what we have already described as the event-nature of a being. The second of these participations, namely in the production of the common world, is what still requires our attention. Merleau-Ponty is endorsing here a *form* of holism: each participant in the world contributes to the overall set of relations within the world both by adding a number of relationships and by altering the quality of others. There is a sense in which each part of the world, through its engaging in relations with the other parts, is irreplaceable and alters the nature of the whole. Each being is, to a degree, irreplaceable because, in substituting one individual for another, the nature of the relations change; there are no *partes extra partes*. This mutual involvement of all the elements is not a mutual causal action of all upon all, a causal nexus, but rather a collective contribution to a common situation. I will attempt an experiential example to demonstrate this point. I do not intend to say that a person *ought* to feel the way I describe about the situation, but merely that such experiences are possible. When walking along a mountain path at twilight, I perceive the mountain as solitary and lonely due to the falling light; the light does not *cause* me to perceive it that way. Rather, the mountain appears as it does through the mutual relation between the mountain, my mood and history, the awareness of the time, the coloration of the light, and so on. The causal mechanisms are distinct, complex, and cannot serve to explain the meaningful perception of the mountain at that specific moment, but my perception of the mountain as solitary and lonely is real and inspires me to behave in particular ways. To complicate and elucidate the structure, add to the scene a moose whom I stumble upon while hiking along the path. The moose changes my prior perception of the environs in various ways; for example, I may feel less alone knowing that I share the place with another fellow being, or I may feel a greater sense of distance in encountering an animal that is very much a wild one. Though I will not speculate on how my presence affects the moose's perceptual field qualitatively, what is clear through the observation of behavior—the startled look, the cessation of activity, the frozen posture, perhaps even the preparation to charge—is that the moose's relation to its surroundings *has* qualitatively changed with my entrance. My presence has altered the moose's behavior. At that point, my perception of the mountain again may be altered, perhaps seeing it as a looming presence, a sentinel, due to my newfound self-apprehension as an interloper, a disturbance to the usual residents of this place; or, if the moose resumes its grazing, I might feel a greater closeness with the denizens of the forest. The point of this example is to demonstrate the continual exchange that occurs between body and world in the constitution of the perceptual field: I am not merely projecting meanings upon beings; the beings themselves come together within the field to correct misapprehensions, intensify feelings, alter the values apportioned to certain things, and so on, all of which occurs at the level of signification. Additionally, with the appearance of the moose, there is a multiplication of perceptual fields, my own and that of the moose, that are distinct and incompossible. How, then, are these two fields connected; how are two divergent fields even possible? If one accepts that the relation with the world is a corporeal relation and that we have knowledge of the world only through our bodies, then the multiplication of fields presents no new difficulty: the fields both open upon the same things, though perhaps articulated differently due to each body's history and distinct sensory composition. Hence our distinct behaviors. Both the others and I open upon things, and we are capable of interaction only through those things and the meaningful institutions that we share. The ability to take part in certain institutions such as language evidences a mutual belongingness (*appartenance*) to a common world of universal feeling, the inhabitation of, or at least recognition of, a common dimension opened within being (*PM*, 194/140).

The emphasis so far has been placed upon the commonality of the universality of feeling, but the notion of participation also requires us to discuss how beings are distinct within this common world. Fields are differentiated with respect to a body's composition and posture within the world, which is meant in two senses. First, bodies are physically composed differently, both between species and within a species, which indicates that "perceptual givens" are not universal and that different perceptions are given to different sensory apparatus. Second, within a specific physical bodily composition, the body in question will live that body in a distinct way. For example, as should be apparent, not all humans will react similarly to the sunset in the mountains or any of the changes in the perceptual field as they were described above; given the unique history of other bodies, the relations between bodies will also be experienced uniquely. These differences, however, take place as divergences from the common background; they are positions taken with respect to the world. While one's corporeality opens upon a common world, individual histories make that world at the same time *one's own*. So as not to imply that there are two distinct worlds at play here, the physical and the historical, it must be emphasized that every physical body, human and non, has a history insofar as it exists within the common world. If bodies are distinct in terms of their histories, and histories are relative to institutions that are all based in turn upon the primordial institution of the body, then the participation in a common world makes possible the various institutions that set beings apart, but history follows essentially from this commonality for the precise reason that beings exist as bodies. This theory of participation removes the problem of whether a being *really* possesses a property by claiming that *all properties are relational*. While some might reply that Plato held the same view—a being has a property by virtue of a relation to a form—it is important to note the difference in the nature of the relation: the being's relationships to other beings determine what properties it possesses. Of course, to say that the being possesses a property is somewhat misleading in this view because the property does not inhere in a substance that is the being. Rather, relations have qualities, and a being just is the relations it enters into. Accordingly, if we were still to say that a being is a substance, we would have to utilize something like Latour's very specific sense of the term explained in the first chapter.

The phenomenon of expression is the means by which beings participate in the common world and therefore allows us a glimpse into a world in which nature and culture are inextricably intertwined. All living bodies participate in a common universality of feeling while at the same time deviating from one another through the myriad institutions established through the body's interaction with that world. While this seems to heal the rift between the two ontological realms of nature and history, it is not yet sufficient to construct the prospect promised in the last chapter. The reason for this insufficiency is that Merleau-Ponty, in his early work, accounts for expression exclusively from the human perspective: the intertwining of nature and history occurs within the lived-experience of a perceiving body. In affirming the unity of nature and culture, then, Merleau-Ponty has done so only for the human order, maintaining the privilege of the living body through the primacy of perception within his philosophy. In order to truly offer an ontology of *Nature* and not of the human signification "nature," these structures must be generalized, must be shown to apply to all beings regardless of the existence of human consciousness. The need for generality is what may have prompted Merleau-Ponty to remark that his early conclusions, correct though they may be, need to be brought to "ontological explicitation," so as to remove them from the realm of psychology.[22] In order to construct our nondominating prospect of nature, then, we will need to examine the ontological structures that Merleau-Ponty was developing at the time of his death in order to see how the order-based metaphysics of his earlier work undergoes a drastic change in sense.

III. THE FLESH OF NATURE

Merleau-Ponty's unfinished classic, *The Visible and the Invisible*, has the stylistic character of commencing anew. In his earlier works, his conclusions are derived from a phenomenological analysis that places a premium upon lived-experience, criticizing various constructed positions for their inability to account for various facets of this experience. Even though structure takes on an integral role in his philosophy, as we have seen, structure remains within a perceiving consciousness and is real in terms of how it is experienced. The operant conception of experience in those texts is that of Husserlian lived-experience, which we had cause to briefly describe in the last chapter. However, with the increasingly vital role that institution plays within Merleau-Ponty's philosophy, another conception of experience comes to replace it. Merleau-Ponty does not name his discovered sense of experience, choosing instead to maintain the older signifier while expressing its new sense by calling for a renewed experience of a perceiver's "brute contact" with the world. Echoing Heidegger's criticism, Merleau-Ponty rejects lived-experience as fundamental because lived-experience is possible only in an already linguistic world, a world that is already articulated and full of established meaning, because the transcendental reduction that isolates lived-experience is something *made* in language (*VI*, 223/171). Given the account of expression offered in the previous section, the fact that singular lived-experiences can be discriminated only through the intervention of language becomes problematic for the following reason: if the transcendental reduction that individuates lived-experiences requires language, and language never *directly* names an object for consciousness, then lived-experience cannot serve as a foundation for philosophy because it is itself a cultural construction. If the foundation for philosophy is still to be sought in bodily experience, which is not a question up for debate for Merleau-Ponty, then the foundation lies in making clear the institutions, intended here in the broadest sense, that condition experience. In other words, Merleau-Ponty wants to arrive at the point at which *Ereignis* is taken to be the proper foundation of and the object for philosophical inquiry.

Ereignis, recall from the last chapter, is the projecting open of being in light of which beings are disclosed through a concurrent unconcealment and concealment such that they manifest as the beings that they are. For Heidegger, recall, conceptions of nature came to dominate the metaphysical tradition because of a failure to differentiate the being of beings, of which nature is a viable denomination, and being as such, the projecting open of a field in which those beings are disclosed. *Ereignis*, then, is itself no-thing, but is the condition for the apprehension of beings as beings through the momentary settlement of the strife between earth and world.[23] *Ereignis* is also, however, something that happens in Dasein, and so Dasein is said to be the bearer of nothingness, the emptiness of the fourfold. As Merleau-Ponty points out, if being is something projected open by Dasein, then nothingness, for Heidegger, is the necessary consequence of the projecting open of being: nothingness, as absence, makes possible the presence of a being as a being.[24] Being and nothingness are not opposed to each other, as Sartre would have it, but rather always accompany each other in the openness of Dasein to the world. The nothingness of Dasein opens upon beings through the projecting open of being. From here, Merleau-Ponty goes on to situate the significance of Heidegger's recognition of the inseparability of being and nothingness in the newfound awareness of how all appearances manifest in the order of beings. Being can therefore never be spoken of from appearances alone because concealment conceals itself in disclosure. To speak of being requires that the concealment no longer dissimulate itself (*NC*, 119). On this point, Merleau-Ponty's understanding accords with the one offered in the previous chapter, though he relies on texts available to him in his lifetime that we have not discussed, but Merleau-Ponty reveals an issue that arises from Heidegger's effort to rethink being as a "moving relation between being and a being, a relation that cannot be fixed" (120).[25] The issue is that if there is a relation between humanity (as Dasein) and being such that being is disclosed only through Dasein's emptiness or openness to beings, then beings and Dasein are "closed up" in their relation to one another (122). In other words, in Heidegger's endeavor to save being as such from being conflated with the being of beings, he appears to shut us up in a hermetic circle of meaningful relations to the world that confronts us, which in turn is shut up in Dasein's projecting open. Here we can see Merleau-Ponty agreeing with Latour's assessment of the limitations of Heidegger's philosophy. Merleau-Ponty, however, does not regard this as an insurmountable flaw of Heidegger's philosophy, a failure that ought to be avoided. Rather, when taken together with the structures at play within the phenomenon of expression, Heidegger's call for *thinking* becomes what Merleau-Ponty will call an "indirect ontology." Merleau-Ponty finds in the circle of relations the need for a parallel investigation to which Heidegger will turn in his later texts: while Heidegger commences with humanity in order to speak of being—"from time to being"—the analytic of Dasein demonstrates the necessity of the complementary investigation that commences with being in order to reach humanity—"from being to time" (122). So Merleau-Ponty is suggesting that the starting point for our philosophical inquiry should be an interrogation of the prethetic world in which we find ourselves and then proceed to explain the specific kind of being that we ourselves are. For this reason, Merleau-Ponty continually emphasizes the importance of an ontology of nature in his working notes to *The Visible and the Invisible*. This move is a risky one, for, as Heidegger shows throughout the history of being, nature has always been the metaphysical concept par excellence, and, if one is seeking to *think* rather than to construct a metaphysical system, one must eschew traditional notions of nature defined in terms of substance. But Merleau-Ponty is not unaware of this difficulty in speaking of nature and actively undertakes to determine nature outside all notions of substance.

The motivation behind incorporating nature into being is Merleau-Ponty's own conviction with respect to being and nothingness. Accepting from early on the Heideggerian position that "there are denials that affirm" (*PM*, 43/30), Merleau-Ponty seeks to express what has been affirmed in the denial rather than remaining content with the determination of beings as the "negation of a negation." Citing Spinoza's declaration that "all determination is negation" (*N*, 208n1/302), Merleau-Ponty proceeds to find this definition of being in Descartes, Hegel, and Bergson as well: absence enters presence through a negation of its own negation (208/156). In other words, beings become present through a negation of their nonexistence. Heidegger makes clear what is wrong with this understanding of being: if being is understood only through its determinations, then one fails to distinguish being from the being of beings for the reason that determinations occur only within a given projecting open of being. Heidegger proceeds to show the complicity of being and nothingness, trying to think being itself rather than through its determinations. Hence, being is thought as *Ereignis*, wherein each being bears the relations between earth and world, god and humanity, as projected from the nothingness of Dasein. But Merleau-Ponty finds in the thought of *Ereignis* the remnants of metaphysical thought's equation of being with the "negation of negation" (*NC*, 125) because being belongs to nothingness and exists only within the nothingness of Dasein (104).[26] Put another way: being is nothingness insofar as *Ereignis* occurs in the emptiness of the *Da*, in the field constructed between the various constituents of the fourfold. The world as it appears to us, then, is the negation of this emptiness because Dasein forgets itself as a *Not* due to the withdrawal of being. In thinking being as *Ereignis*, Heidegger integrates being and nothingness such that they are thought together, each acting as the inverse of the other as on a sheet of paper (*feuillet*), and in so doing shows how there is no direct expression possible of being as such because every direct expression dominates beings by fixing being in a name. But for Heidegger, *Ereignis* as we live it is a *settlement of strife*: we experience being as fixed because we are always on a given path of disclosure (*unterwegs*). Merleau-Ponty believes that he can proceed further than Heidegger, but wholly within a Heideggerian project (*NC*, 146–47), by conceiving of nothingness in a distinct way that avoids conflating being and nothingness (*N*, 208/157). We will look at this conception of negation shortly, but first we must conclude the current line of thought. Rather than attempting to say what being is, which, as Heidegger has shown, is impossible given that being is no-thing, philosophy must find another task for itself: "to bring the things themselves to expression from the depths of their silence" (*VI*, 18/4), that is, to bring about new historical relations with beings by showing the myriad ways in which the beings themselves express a sense outside human signification. But since we are in fact human beings, we can demonstrate the self-signification of a being only through our own significations, which is, in a sense, Heidegger's lesson and the source of his "malaise" (*NC*, 147).[27] Merleau-Ponty's hope is that these human signifying systems, because they are themselves inspired by the world, can regrasp their footholds in the world such that we can see through the expression the source of that expression. The implication, again, is not that there is some original text of which language is a translation, for this would be a view based upon the assumption of being as substance. Rather, the claim is that expression brings to the fore how beings are constituted by their relations, and therefore we are able to reshape human relations continually and, in the process, reshape the being of humanity by engaging with our world in different ways. In short, while metaphysics seeks to give a name to being, the new thought tries to express the relations gathered in the signifying nucleus that is a being. If one accepts that the above elucidation of the structure of expression indicates the mutability of both the being of humanity and of things in the world, that is, the mutability of essence, then the project of constructing a prospect upon the natural world that is nondominating amounts to showing how this structure of expression is present in the world itself and not just in the meaningful perception of a human being. In other words, we are not seeking a prospect *upon* nature, but rather a prospect from *within* it.

There is a multitude of evidence that Merleau-Ponty himself began this project in his later life, but what tools did he use to do so? The first and perhaps most philosophically important expression is that of divergence.[28] Much has been said about divergence in Merleau-Ponty's philosophy, and so we will restrict our discussion of it to what is relevant to our discussion of Heidegger and nature.[29] Divergence, we have seen, is presented as a way of thinking about negation and negativity that will replace the metaphysical conceptions of negativity as lack, nonbeing, and emptiness and thereby allow for a determination of being other than as the negation of negation. Divergence is able to play this role by demoting nothingness from the status of an ontological substance, as is nothingness for early Sartre and most of the tradition of metaphysics, and also by not making nothingness the constitutive element of human beings' openness to the world, as does Heidegger. Rather, Merleau-Ponty finds nothingness within a perceptual field, in the absences, concealments, and divergences that *appear* in the perceived world. As for Heidegger, negativity becomes a kind of relation, but, unlike Heidegger, it is a property of a field of relations rather than of humanity. Due to this change, the question "Who are we?" is answered differently: rather than saying that humanity is an abyss, Merleau-Ponty responds that humanity is a specific corporeal-schema that maintains an open circuit with the things it encounters (*N*, 285–86/223). If negativity is found in the world as well as within the hollow of human awareness, the abyss is no longer a nothingness or emptiness constitutive of humanity, but rather an absence that creates depth, the precondition for all presence, that is found in the world as well (*VI*, 106/77). The sense of humanity is thereby altered from an emptiness that beings come to obturate or even an emptiness that is open to the world. Rather, humanity is conceived as a piece of the world that maintains itself through commerce with that world. To call the continued existence of a being "perseverance" or "persistence," however, would be to miss the central insight of Merleau-Ponty's later philosophy, the *developmental* manner in which beings relate to one another and thereby are *perceived* as persevering.

Merleau-Ponty's word for the manner in which beings relate is "flesh." In Merleau-Ponty's thought, flesh is usually taken to refer to the sensible exchanges between at least one sentient being, typically a human, and another being; and the human body–other body exchange is without doubt his predominant means of describing flesh. Before describing flesh more generally, which will enter us into our discussion of nature proper, we can use the sentient flesh of the body to see the exchanges that underlie the more general structure. A perceiving body opens upon a perceptual field, a field that we have said is populated by beings continually engaged in processes of mutual constitution. Because the perceiving body is within that field, it is subjected to constitution from the outside; this is the aseity of the perceiver. But the perceiver also must perceive itself to be distinct from those other beings in order to act within the field, which is the source of the divergence between the sentient and the sensed. There are also divergences between beings within the field, ways of differentiating beings in perception. If beings are "grasped in divergence [*prise à l'écart*]," we discover an aseity in them as well that comprises the process of becoming in the perceptual field already described in *Phenomenology of Perception* (*VI*, 119/88). The immanence of the perceiver within the field of appearances, then, requires that the relations between the perceiver and the perceived be reversible. The reversibility of relations is the reformulation of the circular relation between perceptual consciousness and world through signification that he advanced in his earliest texts. The difference here being that Merleau-Ponty has come to reject the philosophy of consciousness for Heideggerian reasons (cf. 133/99), and so reversibility is no longer simply the character of the relation between a consciousness and the world, but of the relations between affected bodies. This shift is foreshadowed in the emergence of expression as a central concept in his philosophy as described in the last section. The actual exchange between the bodies is "chiasmic" insofar as there are five terms to be considered in every exchange: the visibility of my own body; the visibility of the other body; what is invisible of myself to the other (e.g., my mind as the "other side [*côté*][30] of the body" [307/259]); and what is invisible of the other body (e.g., another human's mind or the photosynthetic process in a plant). The fifth term is that which plays the role of mediation between the two bodies in exchange, a common element that enables the exchange, for "the chiasm binds together as obverse and reverse ensembles unified in advance in the process of [*en voie de*] differentiation" (310/262). Institutions, as we saw in *The Prose of the World*, are means of differentiation, and now we can say that they are means of introducing deviations into experience. Differentiation occurs in advance through a given institution, but that institution also serves to bind together the visible elements to an invisible meaning, to invisible processes, to invisible behaviors. What remains of consciousness is only the openness to affectation and the participation in a common element that will mediate the exchange between bodies. However, because there are always invisible aspects of the relation, the relation is never *completely* reversible, thus maintaining the impossibility of complete unconcealment (297/249); reversibility remains "always imminent and never realized in fact" (191/147). Divergence, then, opens a dimension of existence both in the perceiver through the self-perception as a distinct being (this we call "mind") and in the thing insofar as it resists complete and total unconcealment. In other words, there is always latency in perception, something that must necessarily remain unconscious (240/189).

Now that the basic structure of imminent exchange that we call flesh is before us, we must generalize the structure such that it is not merely associated with a living body. To allow flesh to remain a property of a living body would be to neglect the problem that flesh is introduced in order to address, that of reintegrating being and nothingness within a field such that neither the philosophy of the subject nor the philosophy of the object takes precedence. For, if the flesh relation remains a characteristic of a living body, then the world is in fact not permeated by negativity, and negativity remains, as it was for Heidegger, a property of a living body. In order to broaden our conception of the flesh-relation, then, we must also look to the various ways in which Merleau-Ponty also speaks of flesh other than that of the living body; for example, sometimes he speaks of a "flesh of the world" that is "distinct from my flesh" (*VI*, 309/261), at others a "flesh of time" or a "flesh of language." The variety of instantiations of the term indicates that flesh is actually a larger structure than simply the relations between a human body and the world.[31] Merleau-Ponty addresses just this point in a working note from May 1960, titled "Flesh of the World—Flesh of the body—Being" (297–99/248–51). In this note, Merleau-Ponty explains how the flesh of the body is actually derived from the flesh of the world, which, "with regard to time, space, movement," is described as "segregation, dimensionality, continuation, latency, and overlapping." The flesh of the world *composes* the flesh of the body because the body possesses a separate bodily existence among other bodily beings, its own perceptual dimensions within which it operates, a continuity of experience and of the body itself, the latency we discussed above, and an ability to act in the field opened in perception. In the same way that in *The Structure of Behavior* the lower order did not explain the higher, however, the flesh of the world is not sufficient to explain the flesh of the body because the two are engaged "in a relation of transgression and crossing from one to the other [*enjambement*]." The difference here is that there are no longer hierarchical orders, but relations in relation to other relations or "fields within fields," historically contingent pell-mell ensembles. With the removal of consciousness as the center that unifies and projects the orders upon beings, Merleau-Ponty adopts Heidegger's vision of the abyssal ground and thus the historical nature of truth. But if an abyss is concurrently located in both the perceiver and in the perceived, then there is "a negativity that comes from the world," or, as he puts it in an earlier note from November 1959, "a *natural* negativity, a first institution, always already there" (266/216). There is no difference between the bodies of things and living bodies at the level of their essence: both types of bodies are bodies of flesh. The difference between them lies in their specific kinds of relationship to other beings such that "the flesh of the world is not *self-sensing* as is my flesh," and yet "I call it flesh nonetheless." The reason that things in the world are still flesh is twofold. First, they are sensible, which means they are situated in the same sensible world where the perceiving body also finds itself, regardless of their ability to sense themselves there. Second, each of these beings is, at any given time, engaged in its own process of becoming; "it is a *pregnancy* of possibles," of which the idea of an object or mere thing "is only a partial and second expression" of the thing's actual mode of being. In other words, if the flesh relation is a relation of mutual exchange between beings in a common, sensible world, and every sensible being necessarily contains a latency, then nonsentient beings, things, because they are sensible, possess a latency and are therefore composed of flesh. Nature is this flesh of the world.

In beginning to speak of nature as a specific formation of flesh, we must take a moment to elaborate upon what it means to say that the essence of a nonliving body is the flesh-relation. In a working note from June 1960 (*VI*, 307–8/259–60), Merleau-Ponty provides a more general clarification of his core notion of a body when he says, "My body is to the greatest extent what every thing is: a *dimensional this*." But what does it mean to say that every thing is a "dimensional this"? As we have seen, every body is within being in a distinct manner, and we said that a body exists in the world according to certain dimensions. These dimensions are established according to how the body interacts with its world, that is, how the body affects and is affected by the world, and, since every body does, in fact, interact with other bodies within the common, sensible world, every body is a "dimensional this" in the dual sense of having dimensions of its own and being encountered through the dimensions of other bodies. Nevertheless, Merleau-Ponty creates a distinction in what it means to be a "dimensional this" for a perceiving and a perceived body. All bodies, both living and non exist according to dimensions that are singular to that body. A perceiving body opens before itself a perceptual field within which there are chiasmic relations between the perceiver and other bodies. As we said earlier, dimensions are the institutions that mediate the entry into the field, both bodily (i.e., natural) and cultural (i.e., historical). A living body is a dimensional this in the sense that it both opens up and exists within certain malleable forms within the sensible world; the living body is the "universal standard of measure [*mesurant*]" of the perceived world.[32] On the other hand, a perceived body, a nonliving body, is dimensional only insofar as it acquires latency within the field of a perceiving body. Because the body does not perceive, that is, is not open to auto-affection, the dimensions according to which it exists are those that are instituted through the chiasmic exchange. The perceived thing may be a dimensional this, but it is measured rather than being its own standard of measure.

But is such a distinction possible if flesh is an ontological structure and not a psychologically dependent organization? This is where Abram is correct to suggest that Merleau-Ponty's philosophy as it is left to us requires further development: though Merleau-Ponty understands being as the totality of proliferated fields or, following Heidegger in "The Letter on Humanism," as "a universal dimensionality" (*VI*, 313/265), Merleau-Ponty still restricts the scope of his philosophy to living beings, which is, of course, a broader conception of being than Heidegger's.[33] The restriction is clear, but comes with a qualification that will propel us into a direct discussion of nature. In a working note from 1959, Merleau-Ponty identifies the "focal point [*foyer*]" of his philosophy as the "perceptual field, insofar as it contains everything: nature and history."[34] Rather than leave the issue resolved, however, and leave the perceived being subjected to the perception of a living being, he goes on to clarify that "instead of saying: perceived being and perception, I would do better to say brute or wild being and 'foundation' [*Stiftung*]." This reformulation, perhaps, will be sufficient to enable us to maintain flesh as the fundamental relation between beings without subsuming the flesh of the world to that of a living body. The caveat indicates that rather than understanding perception literally, we ought to take it more generally as affectation, as the openness of a body to affection from its world. In this way we can discern a three-termed system even for nonliving bodies: a surrounding world with all its sensory richness, a body that encounters it, and the mediating dimensions through which that body is sensitive to its environs. If we take this admittedly brief comment in this way, we can see Merleau-Ponty recognizing that what is needed in order to maintain the fundamental independence of natural processes, the flesh of the world, from being something merely posited by a perceiving being is a model closer to what Deleuze will later call the "plane of immanence." Since nonliving beings continually give and receive stimuli in ways that do not require perception in the ordinary sense (e.g., the heat of the sun affects a stone, which in turn helps maintain the warmth of water in a riverbed), it follows that, were there no sensing beings, the flesh of the world would continue to exist in the myriad ways in which nonliving bodies are passive-active with respect to one another.

I am aware that the above understanding of affectivity is not the usual one utilized in phenomenological philosophy. More typically, affects are discussed in terms of the emotions, feelings, and sensations of a conscious body. All these senses are still contained within the word as used here, of course, but what is likely to be resisted is the attribution of such states to nonliving bodies. It may be surprising, then, that I agree with this limitation. My contention that nonhumans, and more specifically nonliving nonhumans, are affective bodies is in no way meant to attribute conscious awareness to those bodies, as Abram does with his animistic views of nature, or even to attribute intentionality to them, as we briefly discussed Plumwood doing in the first chapter. Rather than minding all beings or extending subjectivity to them, no matter how crude a form, the claim here is somewhat more reserved. Describing all bodies as affective and affecting is meant as a way of *reconceiving* the ways in which bodies relate. I do not intend to ascribe new properties to old substances because the whole point of this framework is that there are no substances, in the traditional sense at least, to which those would be essential properties! If, on the contrary, we import Latour's notion of substance described in the first chapter—namely, that a substance is a particular denomination of a multiplicity of relations that display a regularity of manifestation, allegiances, and powers—then my description of nonliving bodies (that is, specific substances) as affective should be rather uncontentious. If substances are simply a multiplicity of relations, and relations are all ways of affecting other bodies, then a substance *just is* its affects. Consequently, calling nature flesh entails that it is a contingently formed, behaving ensemble of affective relations. This last claim should become more plausible once I explain Merleau-Ponty's assimilation of process philosophy in the following section.

Up to this point we have described many ontological structures, but what remains to be seen is how they hold together as nature. Earlier, nature was equated with the flesh of the world, but, before that it, was said to be the soil that carries us, and, prior to that, the "noninstituted." For what remains of this chapter, we will devote ourselves to understanding how these ontological structures that have been put forward help us to arrive at a conception of nature that is not a system in the sense that Heidegger criticizes, nor a preestablished order such as Latour would reject. What will save Merleau-Ponty's conception of nature from these pitfalls is his discovery of process metaphysics and, specifically, Alfred North Whitehead's concept of "passage."

IV. A PROSPECT FROM WITHIN NATURE

Merleau-Ponty takes the notion of passage directly from Whitehead's philosophy, which he presents as a "new view of nature" (*N*, 152/112) that actively attempts to incorporate the insights of quantum mechanics regarding space and time. While Whitehead's philosophy is known more popularly as process philosophy rather than as a study of the "passage of nature,"

Whitehead in fact equates process and passage insofar as they are both meant to indicate the temporal and spatial existence of natural phenomena as events.[35] For clarity, Whitehead defines an event as "the specific character of a place through a period of time," but which can always be resituated as related to other events within a structure or complex of events (*CN*, 52). For Whitehead, then, what is perceived is a fact, or "the undifferentiated terminus of sense-awareness" (*CN*, 13), and facts are always events that are then abstracted into partial events or factors (15). Considered individually, he calls these factors "entities," but he intends by the term what we have been calling a "being" throughout this text. Beings, then, all participate in a single event and are articulated or abstracted from the passage of nature by means of perception. In this sense, there is also a connection between what we have been calling beings and the multiplicity of beings Latour called "actants": actants share the event-character of individual beings, which are themselves multiplicities. All this is in accord with what Merleau-Ponty has developed in his own philosophy, which leads us to the conclusion that what he finds in Whitehead's philosophy is a concrete notion of nature as the intertwining of space and time within the event, which is, ultimately, what Heidegger is attempting to describe as well when he speaks of the temporalizing and spatializing essence of Dasein. The first step, then, in concretely developing a prospect of nature from within nature is to determine how space and time occur in an event. Unfortunately, because our main concern is with Merleau-Ponty's philosophy, we cannot develop Whitehead's ideas here in all their richness and must restrict ourselves instead to those elements of his philosophy relevant to the understanding of Merleau-Ponty.[36]

We will take our lead here from Whitehead's own comment that "the determination of the meaning of nature reduces itself principally to the discussion of the character of time and the character of space" (*CN*, 33), both of which are determined through an event. The first step in understanding Whitehead's (and Merleau-Ponty's) views on space and time is to accept their rejection of absolute space and time. For both thinkers, the classical definitions of space as an amorphous container and time as serial instants hold within them ontological prejudices, of which quantum mechanics forces the rejection. Whitehead argues that these classical definitions of space and time result in a "bifurcation of nature," into the nature of which one is aware and the nature that underlies and causes that awareness (31). If time is taken to be an ordered and unidirectional succession of durationless instants, and space is taken to be a bounded or unbounded system of points that exhibit no extension, then there is an order outside that exists independent of events. Since perceived events are relative to a perceiver, the assumption of an absolute space and time can be used to explain the persistence of "objects" when they are not perceived by any living being. If, however, one is to question the interpretation of beings as constantly present, as Heidegger argued, then there is little need for the hypothesis of absolute space and time. Rather, what is needed is a way to explain the coherence of events such that the phenomenon of persistence over time is no longer inexplicable.

Additionally, both Whitehead and Merleau-Ponty argue against the ultimate coherence of the absolute views. Whitehead points out the internal circularity of the definitions: temporal instants are known only as relata within a temporal sequence, but the sequence itself is identified in terms of the instants that it is used to define (*CN*, 33). The case is similar for space: we know of "bare space" only as an abstraction from a perceived space because, lacking extension, it is in principle imperceptible (37). Merleau-Ponty approaches the issue of absolute space through the issue of competing geometries, arguing that there is "not an experience of pure geometry in which we can grasp the structure of space" (*N*, 141/103) because the veracity of geometry always requires a standard of measure, and measures themselves are relative to institutions. Thus, as Whitehead argues as well, there is no standard that one can use to choose between otherwise coherent geometrical systems other than one chosen somewhat arbitrarily on the basis of institutions. If there is no noninstituted measure, space cannot be a being that can be defined independently of the measure used to determine it. What makes geometry possible (and physics in general) is the perception of space. Again, the case is similar with time insofar as the conditions for the measurement of time cannot be divorced from the operant conception of time. In this Merleau-Ponty echoes Bergson's criticisms of the conception of time in physics as offering "successive egocentric views rather than bringing about the philosophical coexistence of the times of the different observers" (*N*, 148–49/109). The challenge for the philosopher who wishes to overcome the bifurcation of nature, which is ostensibly one of the main goals of all phenomenology, is to "deepen the relation that exists between these two unities [of space and time]" (157/116) such that the phenomena that the absolute views address can be accounted for without compromising the insights of the relativistic theory of space and time. Merleau-Ponty sums up Whitehead's position succinctly: the bifurcation of nature is overcome through the recognition that the abstractions of science have their basis in a perceived "here" and "now" and that their validity flows from that very basis (158/117). This position, of course, is identical to the one Merleau-Ponty himself takes with respect to the algorithm in *The Prose of the World*: abstractions are expressions that address perceived phenomena. Thus, for both thinkers, the absolute views are undermined through a return to the perceptual situation that is their basis.

But what replaces these modern conceptions? Whitehead is much more specific in this regard than is Merleau-Ponty, which is not surprising given that Merleau-Ponty outwardly endorses and relies upon Whitehead's theories. Space and time are relativized by similar means, and so we will, in the interest of time, attempt to explain them together so as to best show their connection to Merleau-Ponty's philosophy. In order to see this connection as clearly as possible, we will demonstrate Whitehead's points in terms of an example that Merleau-Ponty himself deploys in "Eye and Mind." I reproduce here the passage in question in its entirety:

> When through the water's thickness I see the tiled bottom of the pool, I do not see it *despite* the water and the reflections; I see it through them and because of them. If there were no distortions, no ripples of sunlight, if I saw, without this flesh, the geometry of the tiles, then I would stop seeing the tiled bottom as it is, where it is, namely, farther away than any identical place. I cannot say that the water itself—the aqueous power, the syrupy and shimmering element—is *in* space; all this is not somewhere else either, but it is not in the pool. It dwells in it, is materialized there, yet it is not contained there; and if I lift my eyes toward the screen of cypresses where the web of reflections plays, I must recognize that the water visits it as well, or at least sends out to it its active and living essence. (*OE*, 70–71/371)

What he describes in the example is what Whitehead would call an event because, like Merleau-Ponty and Heidegger, Whitehead begins with the brute fact that *there is*, the contact of a perceiver with a "fact," within which entities or beings are discerned as "factors." As Merleau-Ponty describes the scene, the pool's tiles are perceived there, in the pool, beneath the water, which is to see them in a place. To be in a place is to be discernible with respect to other discernible beings, which is to say that to be in a place is to be within a field. The tiles are discerned in the field to be "at the bottom" precisely through the distortion of the water. The distortion of the water, however, has a temporal aspect to it: there is a rippling effect, a wavering that can be seen only over a period of time. Thus, every place is perceived as being where it is over a period of time. The place "at the bottom" is discerned only because of the mediating effect of the water's distortion. To speak of the tiles alone is to speak of a partial event of the total event of the field. The same is true not only for the cypress, the water, and the light that animates all these elements, but also for the countless unmentioned factors, such as the sky that surrounds the field and delimits it in some way, a clear sky that allows the light to shimmer upon the water. Each partial event is perceived, however, not as a discrete event, but as participating in the one event that is the poolside scene. In other words, the partial events occur simultaneously within the field, but simultaneity does not mean here that they occur at the same instant or point of time. Rather, simultaneity means that the elements are "components of a duration" (*CN*, 56); the passages of the various elements coincide. A duration is a period of time that has an extension, that is, comprises other durations (58). This idea becomes clearer when one considers the field concretely: a given shimmer of reflected light upon the cypress trees comes and goes in its own period of time, the wavering tile has a greater period than the shimmering because the form of the tile is maintained through the shimmering, and the pool itself has a most enduring period insofar as it remains constant throughout the permutations of light and vacillation of water. Yet these three periods of time (among all the others left unmentioned) occur within the single period of time that is the perceived event as a whole. Thus Whitehead says that each duration "retains temporal thickness" (56) because further partial events can be abstracted from a given partial event. For example, one can calculate the surface tension of the water, which, being a force, would possess its own duration. The scene described, then, has continuity on the basis of the common extension of the factors within the total event, which is in turn a single partial event of a larger event that also has a common extension. An example of this continuity between events might be to consider the difference between the pool being located either next to a tidal estuary or along an overdeveloped strip of coastline: the sounds that carry over into the pool area from the surrounding world beyond the cypress trees from the estuary will give one impression of the poolside, while the sounds of the street will give another. There are two points to be gleaned from this addition to the example: First, that the boundaries between any two events are indeterminate. We discussed this point earlier with respect to trying to isolate the boundaries of a single ecological set of relations. Second, that any given event is simultaneously connected to other events that surround it but are not wholly part of it. This continuity between events is also due to their being perceived in a certain place. Because durations are perceived only in a place and from a specific "here" of the "incarnate mind" (107), events that lie beyond that here are distributed in space with respect to the here. Thus, the position of any factor is tied up with its concurrent place within a given time-system that makes up the duration. Spaces, being tied to duration, come and go as those durations pass. This relation is expressed by Merleau-Ponty's musings on the place of the water: as a transparent element, the water itself is not visible, nor can it be "in" space because space is not a container. The water can be seen, however, in two other factors within the event, the play of light and the wavering of the tile. The water manifests itself in relation to these other two partial events, but is located in neither of them. More simply put, we see the water in the way it reflects light and distorts beings within it, but the water is not in the trees or tiles. Because it interacts with these other events, because we see its passage, we know that the water is in fact in space in the sense of being within the common event, but the water lacks what Whitehead calls "simple location."[37] We do not see the water itself, but rather determine its position with respect to the other phenomena that manifest its presence: the water is not *in* the pool; its position is determined in the relation between the event of the tiles, the event of the reflection, and the event of the pool. To determine the place of the water as "in the pool" would be to make the mistake of confusing objects for events (135), where an object is defined as those elements of nature that instantiate within events, but do not themselves have passage (e.g., the color green, the pool,[38] or an electron) (125).

As a consequence of the relativistic views of space and time advanced above, we are now able to draw several conclusions with respect to what nature itself is like. The first conclusion is that nature is not a concretely delimited domain of space and time; time and space *occur* within nature. Though this inversion of the traditional understanding accords with Heidegger's views regarding the origin of space and time within a being that is open to an environment, it differs insofar as it displaces the source of temporalization and spatialization from an individual *Da* and into the world. At first, this reassertion of nature would seem to go afoul of Heidegger's consistent criticism of metaphysics as defining being as nature and nature as the being of beings. After all, if space and time come from nature, then nature seems to preexist space and time. But Merleau-Ponty explicitly defines nature as "pure passage" (*N*, 163/120), meaning that nature does not exist as a thing or kind of thing from which time and space are extracted, but rather that *nature exists as the temporalization and spatialization that occur through the mutual participation of beings, which are themselves events, in an event.* We can see how Whitehead and Merleau-Ponty are actually agreeing with Heidegger in principle with respect to the problem, but disagreeing on the structure of the remedy. If nature is passage, then it is composed of events. Events spatialize and temporalize with respect to one another, which is to say that if there were no events, there would be neither space, nor time, nor beings. Were there no humans, for Heidegger, space and time as we know them would not exist because there would be no *Da* from which they would be projected open. The same is true for Whitehead and Merleau-Ponty, but only insofar as the emphasis is placed upon the "as we know them": since temporalization and spatialization take place between events, not between the elements of the fourfold, not in *Ereignis*, they would still exist in another way for other perceiving events. This transition is imperative to the nonmodern conception of nature that we are attempting to construct because it displaces the privileged position of humanity with respect to nature. Nature is no longer a *human* signification, but has the source of its own signification within itself, with humans attempting to both read those signs and create their own.[39] Rather than finding significance only within itself, Dasein must also find the sense immanent to nature; Dasein must read the prose of the world and, in reading, be called to express that sense in its own way. Nature is the noninstituted insofar as it creates its own sense independent of human projection and ordering, but at the same time, as we have seen, that sense is dependent completely upon those beings whose durations are contained therein.

This point leads to our second conclusion: while we have been discussing nature as a relation of passage between a single perceiver and the passage of nature, it is necessary to multiply the number of perceivers and hence fields in order to comprehend fully what is meant by the passage of nature.[40] Whitehead indicates this problem himself, though indirectly, while speaking of the difficulties that surround determining the situation of an object within an event. The difficulties, he claims, stem from "the obstinate refusal of philosophers to take seriously the ultimate fact of multiple relations. By a multiple relation I mean a relation which in any concrete instance of its occurrence necessarily involves more than two relata" (*CN*, 150).[41] What Merleau-Ponty finds in Whitehead's insistence upon multiple relations is a basis in nature for the flesh relation that disallows the simple, almost mechanistic hierarchization of reality found within the modernist conception of nature.

Though flesh is frequently described in terms of a term-term relation, it is by no means restricted to such a binary system as the pool example shows so well: related in the event are the perceiving body, the water, the pool, the trees, tiles, and light. While it may not be true for all schools of philosophy that they neglect multiple relations, what Whitehead indicates is the necessity of moving beyond binary relations and making the multiplicity of relations the basis for a relational ontology.[42] The question that arises, however, is what to do in cases where two of the relata are perceivers. If the passage of nature is a perceived process, if events truly are the constituent elements of nature, then each witness to that passage will have a distinct perspective upon that passage. Whitehead, of course, recognizes this multiplication of perspectives as well and says, "It is this possibility which makes the fundamental distinction between the new way of conceiving the universe and the old way" (*CN*, 173). These questions can be addressed only through an interrogation of the phenomenon of intersubjectivity, which Merleau-Ponty has made the foremost question of his philosophy by approaching ontology through the questions of expression, perception, and social existence. Whitehead brings us to the point at which Merleau-Ponty can define nature as "that in which we are, that in which we are mixed . . . it is a mixture, not what we contemplate at a distance" (*N*, 164/121), which in turn makes sense of his earlier description of nature as soil. With intersubjectivity, however, Merleau-Ponty endeavors to complicate and elaborate Whitehead's vision of nature such that history has a richer role within it.

In order to see how nature is historical, it is necessary to recognize the level of sociality that is already present in Whitehead's conception of nature. As pure passage, nature consists of events, events that are articulated by bodies such that only a particular subsection of what is discernible as a whole is discerned in any event. What is discerned, though, are bodies, which are partial events of the total event. Any given event, then, is a composite of bodily-events, just as any body-event is also a composite of smaller events; there is a continual layering of strata at each level of individuation. All the bodies comprised by the event participate in making the event the event that it is. For this reason, the event can be decomposed only into the bodies contained therein, with the resultant loss of sense that accompanies abstraction. Conversely, if any of the bodily-events were removed from the total event, the sense of the total event would be different. Each element of the scene is involved in a flesh relation with the other elements, and without any one element's participation in the event, the event is not the same. From this we can conclude that nature, as an event, is an intrinsically collective phenomenon; it is what Latour calls the "common world."

An event is not to be confused with a being, though beings are events. Because an event is not a being, nature itself cannot be spoken of as a thing. If nature cannot be spoken of as a thing, then the question that remains for us to address is, How are we to speak of and think of nature? To this end, we have elaborated several philosophical structures that pertain to our discussion of nature, but we have yet to discuss how these elements come together in order to form a coherent whole. From Merleau-Ponty's earliest works, we discussed how his philosophy was, very broadly speaking, ecological due to its recognition of the overlapping relations between systems and his refusal to reduce those relations to mechanistic causality or render them concrete by means of finalistic explanation. Though at that point mechanism and teleology were rejected because each was but a partial truth of a perceptual consciousness that joined together the in-itself and the for-itself, in his later philosophy the flesh-relation, which takes place between affective bodies and not conscious bodies alone, comes to serve the same function of holding the elements together in an ensemble.[43] In order to see more clearly the import of the transition away from a view of the "order" of nature as something discovered in perception, and toward a view of nature as historically contingent and participatory, we need to look at how the notion of matter is replaced by flesh. As should be clear from Whitehead's conception of nature, there are neither foundational events (events beyond which no further abstraction is possible),[44] nor are there container events (events beyond which no further composition is possible); nature simply is the fact that there is something happening. As Isabelle Stengers notes, this led Whitehead's followers to puzzle over how beings hold together and what constitutes an individual being.[45] Entities or beings, recall, are partial events of a total event, and as such their properties are relative to those other events with which they interact in the larger event. For example, the water's reflection on the cypress trees is itself a partial event that would be manifest differently if the cypress were, say, a concrete wall of an indoor pool that would lend the reflection a greater stability. Where is the substance or matter that would bear the divergent properties? In the light itself? Light is also an event that manifests differently depending upon the other events with which it interacts. Do we then look to the particles of light? Or should we rather say waves? The idea of matter begins with the conflation of an entity, as an abstraction from an event, with a metaphysical substance. What the notion of passage is intended to question is whether or not we can account for alteration in properties without the hypothesis of matter.[46]

The relational theory of space and time claims to do just that, and the flesh is Merleau-Ponty's own vision of how these relations take hold within experience. As most commentators point out, flesh is not matter, but I would add that it is the relational counterpart to matter, meaning that flesh is meant to do the work of matter without being posited as a self-identical substance. We must be careful in saying this, because flesh does not exist as a being either; it is a characteristic of bodily existence that accounts for changes in properties in a relational system without the attribution of substance. The insistence upon this point rests upon Merleau-Ponty's rather strong rejection of the association of flesh with matter, calling flesh a "general thing" or an "element" in the Greek sense: a kind of being that has its own style, behavior, and characteristic gestures and not the static properties of substance (*VI*, 181–82/139). Merleau-Ponty, then, guides us toward what John Sallis calls a "sense of the elemental."[47] Sallis is helpful in that he describes in greater detail than does Merleau-Ponty the elemental as a "third kind" that is neither wholly sensible nor wholly intelligible; it is somewhere in between (*FI*, 180). By "between" here, I do not mean to imply that there are two separate realms of the sensible and the intelligible and that beings exist in some sort of domain that is the amalgam of those other two realms, akin to how Plato conceived of the sensible as somewhere between being and nonbeing. Rather, the mixture is all that there is: all events are elemental insofar as they possess a sense that is generated from the relation of the factors of the event. For Sallis, the elemental provides a new sense to sensibility such that the "sensible" is not devoid of sense, but rather possesses its own "directedness" derived from structures (156). Flesh as an element, conceived in this sense, is precisely that which relates everything while at the same time providing all beings with a singularity of manifestation. In terms of our discussion, Sallis's most important contribution is the bringing together of the perceptual time of human being with that of the passage of nature, conceiving of time as the "interlacement" of phenomena in their self-showing (188–89), which is to agree with the above analysis of Whitehead. Beings show themselves to one another in myriad ways, and time is a measure of the relation between beings in their interdependent manifestation. We said earlier that time is the interconnection of a number of durations in an event, but with the introduction of the elemental, we can see why time does not require a living being's perception. The elemental dimension reveals that there are both passages that occur that are imperceptible to perceivers and passages that occur outside the presence of perceiving beings. But because nonconscious beings participate in the event by supplying a diverse array of durations to it, a general affectivity comes to substitute for the perception of a living body. Here we return to the universality of feeling, but it is now a *true* universality: all corporeal beings affectively participate in the flesh of the world. The elemental allows us to conceive of a completely inhuman time, a time that we can rejoin and, in so doing, alter our participation in it.

Thought of in this way, the elemental opens up a different way of thinking about natural processes and what precisely it means to live within them. Recognizing the temporality of nature opens new possibilities for thinking with respect to what is permanent and transient and then extending this to how one relates to the natural world. If there is no matter, but only flesh, discussions of the "preservation" of nature become rather dubious, as there are no permanent properties or configurations of substance that are in need of protection. Nature will continue to pass; there will always be events, regardless of the devastation that humanity might wreak upon the integrity of relations that surround it in the common world. Are there consequences to such irresponsible behavior? Certainly, as we are currently learning from witnessing the qualitative alterations in the earth's natural systems due to the pushing of those systems beyond certain quantitative thresholds. Additionally, we can say with Heidegger that if humanity is constituted through our relations to other beings, we alter ourselves and our own possibilities in the world when we devastate nature. Last, we can draw on Latour's insight that once we remove the constraints the modernist conception of nature imposes, we are left with only contingent, politically negotiated orders in order to inspect the manner in which we are affecting the constitution of the common world for better or worse. Of course, it remains to be seen how such an evaluation can be made, that is, according to what standard it might be made given that nature itself has ceased to function as a standard. This question will be addressed in the conclusion, but we can make some preliminary remarks in this direction by examining more closely the specific "experimental metaphysics" emerging from this account.[48] In order to address the ethical and political questions that arise through a relational conception of nature, we must look rather at how various levels of fields (events) participate in creating nature. This need is the motivation behind introducing a theory of intersubjectivity, which, with the introduction of flesh into Merleau-Ponty's philosophy, transforms into a theory of "intercorporeity" (*VI*, 183/141). If we look at nature as itself engaged in a dialectic with history rather than as a realm outside and beneath the dialectic of history, it may then be possible to develop better ways of living in nature.

V. THE DIALECTIC OF NATURE AND HISTORY

Now that we have developed a prospect from within nature that does not dominate beings but rather maintains our brute contact with a world that confronts us, we must conclude this chapter by discussing how human relations with nature, and perhaps other living beings as well, might be altered by this conception of nature. But already we must be careful because, in speaking of human relations to nature, we run the risk of reifying the pure passage that is nature into a substantial entity or of considering humanity as something defined against nature.[49] Rather than speaking of the human relation to nature, then, we must speak of behavior toward the common world within which we are mixed. Earlier, we defined behavior in terms of circuits of action that were enabled through the vital projection of forms upon an unformed nature. With the transition from form to flesh as the means by which nature coheres, however, we finally need to deliver on the promise made in the very first section of this chapter: we must view the passage of nature as a behavior. Behavior, though, is dialectical, meaning that there is an exchange between two divergent positions. In our previous discussion, behavior was understood in terms of a body's perception of its situation through forms and the possibilities for action perceived through the recognition of forms. Throughout this chapter, we have been trying to develop a sense of flesh that would involve all embodied beings, conscious or not, such that a body could be said to behave insofar as when it is affected there is a spectrum of possible responses. Consider the existence of "chaotic systems," with respect to which scientists even use the word "behavior" to describe the systems' movements. A chaotic system is one in which the relations within the system exhibit a degree of regularity, describable by algorithms of varying complexity, but at the same time the system is extremely sensitive to changes in initial conditions such that minute changes can result in dramatically different behaviors within the system. Also, they tend to be such that the "initial" point of the system is extremely difficult, if not impossible, to isolate with any degree of precision, making predictions beyond a brief duration highly unreliable. A hurricane is a good example of a chaotic system: its movements are regular, but also nonlinear. Scientists have the means of predicting the movement of such weather patterns probabilistically, but there is no way of doing so deterministically because of the number of variables in play at any given time. Having become deceptively good at determining the course of these systems, it becomes easier and easier to forget that the predictions are probabilistic and confuse this localized ability (e.g., the hurricane will touch land somewhere on the Gulf Coast) for a comprehensive understanding (e.g., the ability to predict in advance of the season the formation of a hurricane, its course, and its strength). Some may argue that given time and technological advancement, humanity will one day be able to compute all the variables and predict these systems deterministically, but this Laplacean view is impossible to maintain once the sciences are freed from nature. What is relevant here is not, however, the human ability to predict, but that, at any given moment of the duration of the event of the system, there are both coherent and divergent possibilities for action *as the system responds to new conditions*.

If one accepts that the passage of nature exhibits behavior and not mechanistic function, an alternative difficulty presents itself due to the adoption of dialectic as a mode of describing behavior, namely, we must avoid turning from mechanism toward finalism. The allure of dialectic is that it is a "situated thought, thought in contact with being" (*VI*, 124/92), that is "wholly subjugated to its content" (121/90), both latent and manifest, such that it "makes its own way, and thus proves that the way is constructable [*faisable*] in the making" (121/90). By remaining in contact with being, the content of dialectic is constantly in transition, that is, the content of the dialectic passes. The passage of dialectic takes the form of history (123/91), which is why Merleau-Ponty's book on the philosophy of history is called *The Adventures of the Dialectic*.[50] Merleau-Ponty does not understand dialectic in a Hegelian fashion. For Merleau-Ponty, the Hegelian dialectic is a "bad" dialectic because its movement is circular and closed insofar as it moves toward an end of history. A bad dialectic is a dialectic that believes there is a position outside the dialectic from which the dialectic could "sum itself up [*faire son propre total*]," without any remainder, in an established doctrine (125/93). The allusion to accountancy is no mistake: throughout *The Adventures of the Dialectic*, Merleau-Ponty proposes that historical change be gauged by means of drawing up a provisional *bilan*, a balance sheet or other quantitative assessment (cf. *AD*, 311/225), that appraises actions in terms of their current value. In making its own total, bad dialectic closes itself off from further transitions in content rather than keeping the running tally of the balance sheet that is formed day to day (*AD*, 274–75/197–98). Bad dialectic considers itself complete by placing itself outside history and gives value to events based upon their progression toward the end, how much they contribute to the total. By stepping outside history's dialectic, even historical events like the Reign of Terror in the French Revolution can ultimately be justified and given a positive role in history. Additionally, in removing itself from its subordination to its content in order to pursue a yet to be realized end, bad dialectic "autonomizes itself [*s'autonomiser*]"; it ceases to respond to its situation in order to seek its meaning beyond the situation today (*VI*, 127/94).

Against this view, Merleau-Ponty envisions the movement of dialectic through the logic of expression and institution discussed above. Here Merleau-Ponty relies upon his notion of divergence to distance himself from the Hegelian dialectic in favor of a "good" dialectic that opens up a "landscape of praxis" (*AD*, 276/199). A good dialectic, because it proceeds by divergence rather than by absolute negation, can proceed only by addressing the conditions it confronts as a result of its prior responses (276–77/199). This responsive dialectic is a good dialectic because it "envisages without restriction the plurality of relations" (*VI*, 127/94) and admits the possibility of a false step through its constant self-criticism (127/94).[51] A dialectic can err, and for that reason its "adventures" are errors, some of which are good and others not (*AD*, 283/204).[52] Such errors can be evaluated only by means of the difficulties to which they give rise, and progress in history is defined not as continual improvement, but as "shaking off false evidences" (*VI*, 124/92): the dialectical passage of history itself explains the errors within it and allows us to correct our "fundamental choices" that led to the current situation (*AD*, 35–36/22). The reference of "fundamental choice" alludes to Max Weber's concept of *Wahlverwandtschaft*, translated into English as "elective affinity" and into French as "parenté de choix."[53] Merleau-Ponty explains the idea of a chosen kinship as an alliance of two distinct institutional sets of relations (Weber's famous example, which Merleau-Ponty also discusses, is that between Calvinism and capitalism), such that the relations of the two become mutually reinforcing and redirect history. Though there are "sketched significations" that might show the consequences of specific alliances, there are always unforeseen consequences to dialectical progression in the same way that the constancy of conditions in a chaotic system cannot be assumed. Since "history is not a rational sequence [*un raisonnement suivi*]" (*AD*, 38–39/24), it is only through erring that we come to know that "certain solutions to historical tensions are impossible" (36/22). Of course, having erred, these errors are part of our historical past, and, with hope, the same errors can be avoided in the future by responding to a situation in a different way, learning new behaviors, rather than pursuing courses that have failed before. History teaches us what errors to avoid, but will inevitably result in further errors.

By proceeding through institutions, history is expressive: it is the "advent of sense" (*AD*, 29/17). The various divergences in our historical endowment are continually giving rise to new possibilities for action, not all of which are necessarily desirable. If universal history, that is, teleological, dialectical conceptions of history, provide meaning to events from the perspective of history's end, hyperdialectic, which is Merleau-Ponty's term for a good dialectic, generates its own sense, meaning that, in the terms developed above, history *speaks*, is expressive. If history is expressive, it must confront a prior ground of meaning that it takes as its point of departure. Clearly, when we speak of history, its prior ground is the past. Dialectically, then, history possesses a double direction. On the one hand, it is a movement toward the future, a future that is a field of possibilities and within which new problems will arise; this is what is meant by "advent." On the other hand, historical change possesses an "inertia" (*AD*, 305/221), an inertia arising from the momentum of the past, a past that can never be erased and continues to inform the current landscape of praxis, if only negatively as a sign of activities to avoid; this is the cultural endowment. In this last idea, we see Merleau-Ponty agreeing with Latour that the linear conception of time in modernity is limited insofar as it fails to account for this thickness of history, the way in which the past never really is past. Divergence within an institution does not result in something other than the original institution; it results in a new sense within the original institution. The novelty of the new position is derived from rearranging the relations within the past institution, not from founding a completely new institution, because even then the "revolutionary" impulse bears a relation to the prior institution.

But our concern in this work is not history insofar as it is human history, but nature insofar as it is historical. What we learn from distinguishing the two forms of dialectical movement with respect to this latter concern becomes more apparent when we reintroduce the notion of behavior in nature. If nature behaves, then it passes by means of a dialectical movement, but the movement of a good dialectic that comprises a plurality of relations and not simply two opposing relations. Such a dialectic is historical, meaning that nature carries its past within itself in the form of institutions and expresses a sense through those institutions. Nature as pure passage is the flesh of the world, meaning that it is characterized by imminently reversible relations between a plurality of beings. Insofar as those relations that compose nature remain fairly stable, which is to say the relations are not radically reorganized, then the institutions of nature remain stable. The proportional stability of relations is what allows science to describe the chaotic systems of these manifestations with algorithms. As the name for the institution created through the passage of events, nature does not simply have a history as intimated above; it is a historical phenomenon; it is "the memory of the world" (*N*, 163/120). To be clear, when we call nature historical, we refer to the past events that remain immanent to current events through the instituted behaviors of the beings, from hurricanes to humans. Nature thus presents us with an inertia that is not instituted by humanity; it is a collective phenomenon, a common world in which all beings participate.

VI. CONCLUSION

When we began this chapter, a prospect upon nature was promised that would alleviate the human domination of beings. Have we arrived at that point? Nature, it is true, is no longer considered as a spatial container through which infinitesimally small instants of time parade. It is, rather, taken as a collective phenomenon, an institution formed through the collective passage of events.[54] Every event comprises durations, and spatiality emerges as a relation between durations. This view led to regarding nature as a historical phenomenon, but not in the sense of being a social construction. Nature is the past that is borne within the present moment, is the institution that provides direction to physical processes. As historical, nature behaves, and its behaviors have a sense with respect to the institution that it is and will continue to be. What changes is not the institution, though new ones are always possible, but rather the sense of the institution.

As an institution, nature comprises events, events that are the envelopment phenomena of the beings contained within their field. Here we have the second inversion of the traditional view: rather than beings receiving their meaning from nature, nature receives its meaning from the relations between the beings involved. What makes this view so provocative is its basis in the theory of flesh, which gives primacy neither to ideas and meaning (the invisible), nor to materiality or sensation (the visible). Rather, flesh, and relational ontology in general, calls our attention to the fact that there are no pure meanings or pure bodies, only meaningful bodies. If nature is the unity of events within which all bodies are mixed, then those bodies, to the degree that they are open to affection from the world, are all witnesses to the passage of nature. The generality of flesh through the generality of affectivity within bodies requires that every body be expressive of a sense, but with that sense measured through the dimensions within which that particular body is capable of dwelling. Nature carries us all because it *is* all of us, and by "us" I do not mean merely human beings, but all those other bodies with which we share the earth.

Every "thing" is connected within nature, but in saying this we must be cautious to avoid the metaphysical systematization that it implies. In accordance with Heidegger, nature does not exist as a being, but we can still see beings as having particular ways of emerging into this common world. There is, however, a major ontological shift that has taken place from the Heideggerian model: the humanistic ideal of "man" as the "shepherd of being," as the site for the projecting open of being as *Ereignis*, has been abandoned. Even though Heidegger opposes every naming of being that would turn being into a thing, in a sense he makes being into a property of Dasein, something that would not exist without humanity. Consider the following statement from *Contributions to Philosophy*: "Being (as *Ereignis*) needs beings so that being may essence. Beings do not need being in the same way" (*BP*, 31/22). What a curious statement if being were not simply a projection of Dasein! In Heidegger's thought, however, the statement is not curious at all because beings would still "be" if there were no humans; they would simply not be beings for the reason that they would not be subject to the projecting open of being. Being is possible only because *there is*. Merleau-Ponty will agree with this last point with respect to the "there is," but not on the more pivotal point that only humanity has language, which is the "house of being." For, as we have seen, language is, for Merleau-Ponty, a particular case of the general expressivity of corporeal existence, it is a dimension within which a body dwells: bodies are expressive and, as expressive, belong to the primordial institution, the primordial history of nature. As deep as his intellectual relationship with Heidegger is, Merleau-Ponty is able to bring being and nothingness back out of humanity in a way that radically challenges humanity's supposed privileged relation to being. Heidegger is still a humanist, even if only a "curious" one, which is a position tied to the metaphysics he critiques (*N*, 176–85/131–38, esp. 184–85/137). Though Heidegger will make various qualifications to his humanism in "Letter on Humanism,"[55] it remains true for him that only humans have a relation to being, even going so far as to distinguish the human body from "an animal organism."[56] While Merleau-Ponty may still try to describe what is unique about human existence, he will never admit such a rupture between animals and humanity because all life is "a power to invent the visible" (*N*, 248/190).

The consequence of his divergence from Heidegger is that being can once again be seen as a system, but a system without any fixed hierarchical relations because all beings participate equally in the common world. This notion has powerful ecological import in terms of humanity's current situation with respect to the environment. If the system of nature is no longer hierarchically stable but merely a name for the past that has led to the current state of an unstable equilibrium among the relations between beings, then nature is free to express a new sense depending on alterations within its instituted being. That statement is rather abstract, however, so let us consider it more concretely. The stability of nature is a historical phenomenon, built up from relations among beings. If we consider the environment of the earth as a partial event of nature, we can abstract from that event all the beings with which we are familiar: the atmosphere, humans, various fellow animals, plants, and so on. If certain relations are pushed beyond specific, perhaps imperceptible, thresholds, there will be a change in nature, meant here in both possible senses. Humanity has for quite some time been pushing nature toward a threshold, but this will result in not "the end of nature," as some have claimed, but rather an expressive recalibration in the flesh of the world. I do not intend here to trivialize mass extinction, perhaps even of human beings as we currently know them, but rather I would like to emphasize that even a dramatic change in the climate is something from which nature will emerge unscathed. Nature will, in an impoverished way, continue to pass, but it will have a new historical architecture. Beings will continue to participate in the common life of nature, its processes and intercorporeity, regardless of humanity. On this view of nature, beings are no longer dominated by the idea of nature; they are "let be" to express their own sense through their elemental existence.

CONCLUSION

NATURE'S NORMS

> Ultimately, a durable relationship between we humans and our planetary partners must be built on the kind of perceptual, epistemic and emotional sensitivities which are best founded on respect, care and love.
>
> —Val Plumwood, *Environmental Culture*

It would be a rather unsatisfying conclusion to this book if, as the end of the preceding chapter suggests, the conception of nature we have taken such pains to develop necessarily led to quietism with respect to what constitutes a *good* common world. After all, the foundation for an environmental ethic should be able to provide some form of guidance with regard to how to behave within that world. The kind of guidance I envision, however, will not be of the *moral* variety, by which I mean that I do not intend to supply a *rule* or *decision procedure* that we could follow somewhat automatically and without reflection, or even specific obligations regarding our interactions within nature. On the contrary, what emerges from our discussion of Latour and Heidegger is a particular *ethical* vision of agency that, when combined with the conception of nature that emerged from our engagement with Merleau-Ponty's ontology of flesh, opens the possibility for us to experience certain kinds of feelings toward various human behaviors. Here, at the conclusion of this line of thought, it proves necessary to show how the preceding has led us to a point where the ideal of mastery might be challenged by embracing nature's mystery.

The first step in revealing how the conception of nature as the flesh of the common world can serve as a foundation for an environmental ethic is to analyze the manner in which it is or is not able to resolve the inherent tension between the two seemingly competing commitments of an ecological ethic. Steven Vogel introduced us to the idea that specific aspects of environmentalism were in conflict with one another. On the one hand, there is the commitment to antihumanism in the sense that the motive for improving the human relationship with the rest of nature should not merely be that it is in the general human interest to do so. On the other hand is the commitment to the independence of nature, the idea that nature should be preserved free from human intervention. At the time, I had promised to develop a conception of nature that would support each of these intuitions without falling prey to the antinomies that Vogel describes emerging from holding them simultaneously, but at this point it should be apparent that this is possible only by transforming the sense of each of the commitments. Let us examine each in turn.

As I just suggested, it may appear that I have failed to develop the sought-after normative foundation, since the last chapter seems to have arrived at a rather anthropocentric position regarding *why* Western consumerist attitudes and industrial practices should change. After all, if the ultimate consequence of the view is that nature will go on with or without humanity because every being participates in the constitution of nature, what reason is there to change our behaviors other than a desire to preserve our own continued existence? This objection may be convincing given a dualistic mind-set because it bifurcates human interests from the well-being of nature. As Plumwood has convincingly argued, however, removing ourselves from the context of a dualistic metaphysics also allows us to eliminate the false dichotomy between human interests and the interests of other species.[1] In her own words:

> Considering your own interests does not imply that you cannot also consider others' interests as well as, or as related to, your own. Prudence does not consist in counting *only* one's own interests as reasons for acting or not acting, . . . but in taking one's own interests into account in a consistent way, and counting injury to them as among your reasons for avoiding an action. The idea of prudence says nothing about consulting your own interests *to the exclusion of others*. That is not prudence any more than it is rationality—it is selfishness, or egocentrism.[2]

This passage is reminiscent of John Stuart Mill's claims in *Utilitarianism* regarding the sympathetic feelings of humankind: we are simply the kind of creature for whom our own interests and the interests of others are hopelessly intertwined, such that our happiness and well-being necessarily involve the happiness and well-being of others.[3] The problem with traditional anthropocentrism is its human-*centeredness*, the manner in which it systematically either ignores the interests of the others with whom we share the planet or prioritizes human interests over all others. As Latour showed through his conception of an actant, agency is comprehensible only as a collective phenomenon, and so neglecting the interests of the nonhumans that contribute *essentially* to agency is decidedly imprudent. Prudence, if separated from an imprudent human chauvinism, can then be an integral element of an environmental ethic: recognizing how the various interests of humans and nonhumans are intertwined, it is prudent to strive to produce the best common world. Calling the environmentalist's commitment "antihumanism," then, overstates the case. What I commit myself to here is better thought of in terms of what Plumwood calls throughout her work an "anti-centrism." In being anticentric, one can make sense of situations in which deferring or putting aside one's egocentric interests can actually be prudential insofar as one's self-in-relation to others is enriched.

Regarding the second commitment to nature's independence, the impact of the hermeneutical shift in our conception of nature is a bit more dramatic. Consequently, it will be helpful to recapitulate some of the ground we have covered in order to see in what way there remains room for the independence of nature.

If one thing is clear from the preceding arguments, it should be that we should not understand the freedom of nature dualistically to mean that human presence should be effaced within nature in order to preserve it. Both Latour and Heidegger have offered reasons for us to be highly suspicious of the tendency within environmentalism to regard human presence as inimical to nature. In Latour's case, we saw that the suspicion arises from a rejection of the practices of purification within the modern constitution. Nature and society, the two poles of the traditional dualism within Western philosophy, are the two "collectors" within modernity, that is, the two possible ontological statuses that a phenomenon can have. While all beings are thought to originate in one domain or the other, there is a simultaneous proliferation of hybrid beings, phenomena that cannot be realistically explained by the agency of either of the poles separately, that the hyperseparation of nature and culture makes possible. The dualistic ontology, even if it becomes naturalistic or constructivist by reducing one collector to the other, is a bad way of understanding the world because everything that exists, exists as a multiplicity of diverse relations. If nature is rejected as a substantial domain ordered in advance according to its own internal principles, then human interaction with nonhuman elements cannot be condemned strictly on the basis of humanity's violating that fantastical order. For Heidegger, the reasoning is somewhat different, but we arrive at the same conclusion: there is no immanently ordered domain of nature. His rationale for this conclusion lay in the history of being. Specifically, nature is a conceptual category that emerged as an externalization of the explanations for why a being emerges into the world in the manner that it does. As *physis* transforms from an inner principle into an external domain, the ideal of mastery becomes more firmly entrenched insofar as the nature of nature becomes ever more unquestionable. The conception of nature as an energetic system becomes so ingrained that the mere suggestion that a hermeneutic project oriented toward reinterpreting the concept is possible is liable to be met with a high degree of skepticism. Regardless of this point, Heidegger's critical engagement with the history of Western philosophy directs us toward the way in which liberating beings from this mastery is not a matter of disengaging from them, but rather that beings must be engaged with in such a way that we remain open to their mystery.

With Latour and Heidegger's joint agreement that disengagement from other beings is neither possible nor desirable as an ethical strategy, the question arises as to what these desirable modes of engagement are like. From Latour, we saw that the impossible project of mastery within modernity is to be replaced with a new constitution that explicitly politicizes the relations between beings. The nonmodern constitution functions with a complex set of recursive relations between various "houses," each of which performs a specific function. The sciences govern entry into the collective, evaluating whether various propositions adequately capture a set of relations observed in the world. The political process then sets about establishing a provisional cohesion of the collective given the propositions left outside the collective as well as with respect to the interests of the collective's various constituents. Economists, whom we have unfortunately overlooked to this point in the interests of our specific focus on nature, provide a common language to express and assess the inner workings of the collective and the impact of its current externalizations. Last, the moralists then evaluate the works of the other three functions, suggesting ways in which the work of collecting needs to be resumed and reworked. In light of this last function, the earlier mild criticism of Latour should be revisited: while mediation has always been with us in some form, the history of mediation could be conceived in terms of how the work of collection is performed in any given historical period. In this way, mediation is a parallel concept to Heidegger's being and is aligned with the understanding of institution in Melerau-Ponty's work. Regardless of whether one accepts that last point, giving nonhumans a voice at all the various levels of the constitution reintegrates them into the political work and acknowledges the integral role that nonhumans play in the formation of actants. Nonhumans, then, are to be engaged with as fellow citizens in this political activity of constituting the common world. In Heidegger's case, the recommendation for improved relations to the nonhuman world comes through his injunction to "let beings be," to remain responsive to the innumerable ways in which a being can disclose itself and thereby allow it to retain its independence from the disclosure of Dasein. If the devastation of the earth is sanctioned and perpetuated via the domination of beings within a single projecting open of the being of beings as constant presence, then in letting beings be Dasein ceases to master them by confronting the mystery while quashing the quixotic quest to dispel it. Heidegger's proposal was that a mindfulness of the *Ereignis* structure, how being is projected open, might then enable the development of another, truly alternative conception of nature that would no longer rely on the substance-based metaphysics. If, as Heidegger argues, the metaphysical tradition's assumption of nature as a substantial domain is what has sanctioned the devastation of the earth due to its status as a resource for human use, then a new relationship with the earth will require a nonsubstantial conception of nature.

For both philosophers, the independence of nature is conceived not as a radical separation from humanity, but as a kind of political independence from the domination or mastery of humanity. In both cases, that mastery is of a hermeneutic nature, wherein the operant *conceptions* of nature in Western thought both enable certain exploitative dispositions and preclude other more charitable and egalitarian dispositions. For example, the notion that one could demonstrate *compassion* toward nature is, for most, a far-fetched idea simply because nature is not considered to be capable of suffering. While I would hesitate to personify nature in order to establish that it suffers as a result of certain human (and nonhuman!) behaviors, once the interpretation of nature the project of mastery necessitates is relinquished, it is at least *possible* to think of what nature's suffering is like. We will return to this idea shortly, but given that both Heidegger and Latour recommend that nature be left behind as an ontological category, we must first clarify the conception of nature that makes the consideration of nature possible.

In order to create a relational conception of nature that is not based in substantial metaphysics, we extracted certain constraints that a nonmodern conception of nature needed to observe from both Latour's and Heidegger's philosophies. Both reject the notion that nature is a being or a thing. They also both reject that there is a necessary order to nature. While there are important differences between their reasons for these rejections, I am overlooking those for the moment in order to draw attention to these points of connection in terms of the current discussion. Latour's positive contribution to the nonmodern conception of nature was that if nature were going to be said to exist, it would have to be as a name for an actant. Since Latour does not explicitly explore this possibility, Heidegger's philosophy was somewhat more helpful in directing us toward a nonmodern conception of nature.[4] His main contribution to our investigation in this regard was to show *how* the substantial view of nature might be replaced. To this end, we looked at how the intuition that led to the supposition of a substantial nature, namely, how the being of beings that was initially conceived in Greek philosophy as constant presence *conjoined* with the absencing that accompanies any presence comes to be exclusively identified with constant presence, is affirmed in his concept of "earth." Earth is what resists complete disclosure *in a being*, that is, what remains concealed in any disclosure of that being. Earth, then, retains the original Greek intuition that total disclosure is impossible without necessitating a substantial nature to explain presence. If every being is of the earth, then every being will, in some way, lie outside human attempts to order and systematize the world: there will always be something more that lies outside any disclosure of that being. Since the being disclosed will always remain mysterious, that is, its otherness can never be overcome, there is no single disclosure that will capture the essence of said being.

While Heidegger's philosophy gives power back to the beings encountered and actively attempts to move away from the conflation of being and the being of beings, it nevertheless has the consequence of making the view of nature relative to humanity. *Ereignis* occurs in the *Da* of Dasein; animals are poor in world, and nonliving nonhumans are worldless. There is, then, a tension internal to Heidegger's view where, on the one hand, his relational ontology is able to empower beings by placing aspects of their existence outside human mastery, while, on the other hand, he continues to argue for humanity's possessing a privileged relation to being. If one were to be more sympathetic to Heidegger along this line and argue that his views concerning human distinctiveness are actually aimed at discrediting the modern teleological conception of the organism, one would still have to admit that his philosophy is committed to the view that language is the sole province of humanity, and thus that humans alone have access to being as such.[5] The tension resides, then, between the concurrent antihumanistic assault upon humanity's privileged position as the orderer and master of beings and the humanistic affirmation of humanity as the shepherd of being, the being before whom an order is projected, the being who gathers other beings into a meaning.

For this reason, I directed attention to Merleau-Ponty's philosophy as a way of developing a relational ontology on a phenomenological foundation. By tracing the transitions in Merleau-Ponty's views, we saw more clearly the motivation behind pushing Heidegger's ontology beyond the limits within which it is comfortable. Rather than continue to speak of Dasein, Merleau-Ponty's insistence upon perception as a fruitful and revelatory starting point for philosophical inquiry and, more specifically, the primacy of *a* body's relatedness to other bodies, eventually led him to a view of nature as the common world from which each body diverges. But to speak this way discounts his equal affirmation of the world as nothing more than the composite of relations between existing bodies. By emphasizing the role that Whitehead's process philosophy plays in shaping Merleau-Ponty's late views on nature, I argued for a generalized conception of flesh such that it is no longer based upon human perception and embodiment but is, rather, a characteristic structure of relation in general. Under the more general, process-oriented interpretation of flesh, space and time become the indissociable products of the relations between bodies. Because the flesh-relation comprises both the production of excess and the retention of latency, Merleau-Ponty is able to maintain Heidegger's insight concerning the mystery, while at the same time displacing humanity from its privileged position as the sole dweller in the house of being, the only linguistic being. In Merleau-Ponty's view, bodies are linguistically relevant, not humanity alone, though bodies might speak diverse tongues. Nature, as the common world, is expressive of a sense that can never wholly be articulated from any one being's perspective without loss of the mystery.

With the discovery of primordial expression within each body, every body can be said to contribute to the common life of nature. Space and time are formed through the connections and relations between events. Bodies communicate with one another through their mutual affectations, forming behaviors through those historical relations. In Merleau-Ponty's view, being, which can be defined as a general affectivity, is a stratified phenomenon, approachable only from within according to particular dimensions. Being is the universal dimensionality, an enveloping phenomenon of *what is*, not another word for "nature." Nature is only one strata (*couche*) in this stratified being. In order to see how the layers go together, to see how nature is the common life of all, one must practice what Merleau-Ponty calls, in a working note from June 1, 1960, "transcendental geology" (*VI*, 306–7/258–59).[6] Recall that Merleau-Ponty will admit only of an indirect ontology, an *expression* of being rather than a naming of it, which leads us here to say that nature, as the collective life of expressive bodies, can also be "mined" for a sense and a history. Thus, while we have been discussing the being of nature, what we have really been discussing is an indirect expression of being itself, one of the strata of wild being (*VI*, 230/178). Nature, then, must be studied, not as another word for "being" as it was in metaphysics, but as a "leaf [*feuillet*] or strata of total Being" (*N*, 265/204).[7] While *feuillet* is the word for a leaf of paper, it also has a technical meaning within the science of geology: a *feuillet* is a subdivision of a stratum. It should come as no surprise, then, that flesh can be described as a "gangue" (*VI*, 24/9), a vein of earth or rock layered between two geological strata containing ore; flesh holds the strata together with beings appearing in the midst of these relations. Though anachronistic, we might call Merleau-Ponty's later philosophy a "geo-philosophy" that takes nature, life, and history as three strata of wild being held together by relations of flesh.

But the question with which we began this recapitulation concerned what impact the reformulated conception of nature would have upon the commitment to the independence of nature. If nature is just a way of describing the common world of interrelations between bodies, then as long as there are bodies in the world, nature will have a sense other than that given to it by humanity. Accordingly and in addition to the reasons Latour describes well in his work, we must reject the social constructionist position on nature even as we conceive nature as a historical phenomenon. In this sense, nature retains an independence from humanity without engendering an ontological dualism. We can thereby affirm the two competing commitments in an altered form without engendering a contradiction in the way Vogel describes. Antihumanism has been softened, but also strengthened, into a general anticentrism, which is quite compatible with the view that nature's independence lies in the fleshly relations of bodies. There is neither dualism at the ontological level, nor a transcendence of nature such that it exists as a separate being with its own inherent order. In order to make these commitments consistent, however, it was necessary to sacrifice the very thing that provided an environmental ethic with its normativity, an inherent order. If the challenge of this book was to construct a view of nature that would enable ecologically minded philosophers to argue for alterations in human lifestyle while also maintaining that an immanent order to nature cannot serve as a normative foundation for those arguments, what remains to be seen is what kind of foundation for normativity remains in the reformulated process conception.

Of course, to advance a normative framework in its entirety is not a plausible endeavor in the space that remains, but I would like to say some words regarding how I envision the normativity of this conception of nature when taken in concert with the phenomenological project. Recall that the tensions Vogel pointed out between the commitments arose from the attempt to use them to criticize and evaluate human activity as transgressing or going against nature. If humanity is but one of myriad participants in the creation of the common world and there is no one way that common world is supposed to be, it seems as if we may be no better off in criticizing certain human behaviors than before. However, the position I have described, along with the commitment to anticentrism, lends itself to developing normative criteria to address whether actions do or do not promote the domination of nature. Latour's democratic political organization of the collective and the Heideggerian injunction to remain open to the mystery supply dispositions through which it is possible to avoid domination of others. Being dispositionally opposed to dominating other beings within the anticentric framework requires being attuned to several different aspects of a situation: (1) how the various interests of diverse actants are intertwined; (2) the concrete effects upon the common world created by choosing to act upon a certain set of interests over others; (3) how the choice one makes might affect the constitution of a given actant or actants; (4) how interpreting the situation otherwise opens or forecloses possibilities for action; (5) how one's relationship to other beings is transformed by the action; and so on. This list is obviously not intended to be exhaustive, but to provide a sense of what kind of foundation for evaluation it might be appropriate to seek out.

While it may be argued that some of (or all) these aspects of the situation were already a part of our deliberations prior to adopting the disposition of being open to mystery and accepting a more participatory role for nonhumans, it is certainly the case that these aspects were not a central focus to deliberations.

None of what I have just said, however, should impede our ability to engage with other beings within nature in such a way so as to produce the best common world, but I would like to add at this point that some kind of overarching harmony of all beings is simply not possible. As Latour notes, "In abandoning mononaturalism, political ecology does not promise peace. It is only beginning to understand what wars it has to fight and what enemies it has to learn to designate" (*PN*, 219). Rather than pitting subjects (humans) against animals and objects (nonhumans), there emerge a whole host of antagonistic relations within the common world between its members. There will always be adversity within the common world, but the point is to develop structures and dispositions to deal with that adversity without resorting to political or hermeneutical domination.

Focusing on dispositions such as the openness to mystery makes available a set of virtues such as compassion and friendship that were previously thought to be unavailable with respect to nature. But saying that one should be compassionate to the earth is still unhelpful for two reasons: (1) we still lack a standard by which to discriminate a better set of relations within the common world from a worse one; and (2) even armed with such a standard, it is not clear to what extent nature is something that can inspire our compassion or with which we can be friends. In closing, then, I would like to address these last concerns.

As has been a popular tactic within environmental philosophy, focusing on the concept of *place* rather than nature as a whole provides greater normative force to these dispositions.[8] Nature as a whole is composed of places, and, even if some might find it difficult to claim a love for nature, it is hard to imagine a person who would disavow a love for any place at all. Places, as we are speaking of them here, are partial events of nature as a whole. As partial events, they, too, comprise further partial events, bodies. Because the relations between bodies compose places, places will also possess latency, possibilities for further unfurling, as well as a history of how they became the places that they are. Places, like bodies, have an integrity that arises from their history, but they will also change with respect to influences in their surroundings, which is to say that places respond to their environments.

If one accepts that place can be taken in this way, which is a consequence of the relational view of nature, we can then speak of places as viable loci of *care*. But simply saying that one ought to care for places does little to resolve the conceptual difficulty we currently face of defining what actually constitutes a good state of the common world. My final suggestion will be that such a criterion is found in Georges Canguilhem's *The Normal and the Pathological*.[9] To those who know of Canguilhem, this suggestion may seem a bit odd given that his research primarily involves the history of science, medicine, and biology. *The Normal and the Pathological*, in fact, is a prolonged meditation upon those two concepts in the (French) history of medicine, followed by a questioning of the very idea of a norm, disease, and health and whether those ideas can be exported from the medical/biological sciences into the social world. It is this analysis of norms that I will focus on in order to discuss what might be an appropriate sense of normativity emerging from the ontology.

Canguilhem argues for a conception of health as a "silence of the organs" (*NP*, 118): as long as the body, or in our case, the place, sustains itself without limitation in exercising its "normal" behaviors, we can call it healthy. The key is that the idea of a norm, for Canguilhem, is a highly individual one (121), in which there can be anomalies, deviations from the normal state, that are still not considered pathological (136). Norms are relative to situations, and disease itself institutes its own, inferior norm (183). One can call a norm superior to another, for Canguilhem, "when it includes what the latter permits and what it forbids" (182). Disease, then, is a narrowing of the scope of a body's potential for action, but it can also open a "new dimension of life" (185–86), meaning that living according to the diseased norm can actually institute a novel set of relations that might prove to be superior given changes in conditions. There is a distinction, then, between what is normal and what is healthy: good health requires the ability to recover from sickness, while simply behaving according to a norm does not necessarily imply such health (199). Notice, however, that there is an additional *valuation* that takes place here in this distinction: the norm is strictly referring to the stability of instituted relations, while health refers to the ability of the body to achieve a new balance when threatened by external conditions, to respond with new norms of behavior, to be *normative* rather than normal. In this way there is a parallel to be made between the sense of health introduced briefly here and contemporary ecological conceptions of resilience. The difference between them is that resilience focuses on the *reestablishment of the same within certain limits*, whereas a state of health for Canguilhem requires only *reestablishment of some norm* and *the retention of normativity*. Canguilhem's views, of course, are much more nuanced than explained above, but to give them the time and consideration they are due would be to begin the project of setting forth a specific ethic. My purpose in introducing these ideas, however, is just to demonstrate that a normative position is possible based upon the metaphysical considerations introduced here.

Even on the basis of such a limited account, however, we can still see how places can have normativity in Canguilhem's sense of the word, responding to both their external surroundings and changes in internal constitution by establishing new norms, new behaviors. Places, of course, do not "will" themselves to institute new norms, but, then again, neither do our bodies. In the same way that norms are based in situations, what is normal for a place will also be situationally dependent, meaning that not all changes are bad. Changes of which to be wary include those that would limit the ability of the place to respond to further changes, such as how slash-and-burn agriculture in the Amazon eliminates the rain forest's ability either to restore itself or to institute new, superior norms. Deploying Canguilhem's conceptions of the norm, normativity, and health within the context of an ecological ethic would be beneficial for the reason that they do not rely upon a reference to any one kind of being's "needs," "welfare," or "interests"; nor does it ignore those needs, welfares, and interests in general. The merit of shifting focus from nature as a whole to places is that we have a much better understanding of what might be valuable within a given place than what might be valuable in the cosmic scheme. Indeed, it is doubtful that there is anything an individual being with human duration and a human scope of action could do that would be either good or bad at that scale.

At least for me, the writer, one of the truly mysterious aspects of the account offered here is that we ourselves, in a certain sense, *are* the places where we reside. We have no fixed place in nature because our place is simply to participate in the ongoing and continual processes of mutual individuation and coevolution that is nature's very existence. Our behavior affects what is possible within places, and they in turn condition what is possible for us to accomplish alongside our fellow inhabitants of this planet, but our belongingness to nature is utterly non-normative. In this respect, Vogel is right about nature: we cannot appeal to it in order to learn what it is we are supposed to do. Our selves, which grow in relation to the phenomena we encounter, are shaped and molded through both our actions within nature and the narratives we tell about our relationships. While places frequently serve as a setting for those stories, relational ontology calls our attention back to how, even as we remain individuated in so many meaningful senses, we *belong* to places. In forgoing the project of mastery of nature, we liberate ourselves along with the nonhumans. We are no longer called to *sacrifice* in the name of nature, but to transform ourselves in ways that enhance our selves by cultivating opportunities for richer relationships between ourselves and other beings. In addressing Heidegger's question, "Who are we?" in this context, we might reformulate the question and ask, "Who is it we endeavor to become?"

in the name of nature, but to transform ourselves in ways that complicate as well as cultivate opportunities for richer relationships between ourselves and other beings. In addressing Heidegger's question, "Who are we?" in this context, we might reformulate the question and ask, "What have we undertaken to become?"

NOTES

INTRODUCTION: THE QUESTION OF NATURE

1. Bryan G. Norton, *Toward Unity Among Environmentalists* (New York: Oxford University Press, 1991), 5–6.

2. Ibid., 9, 12.

3. John van Buren, "Critical Environmental Hermeneutics," *Environmental Ethics* 17 (1995): 260.

4. One difference between van Buren's categories and my own on this matter is calling the first set of concerns "epistemological." Certainly he is correct to indicate that what is at stake is our way of knowing nature, but for reasons that will become clearer as the argument moves forward, I would prefer to situate the analysis at the metaphysical level. Briefly, the difference could be expressed by saying that the practice of critically evaluating different views of nature is itself the task of metaphysics. Ibid., 261.

5. Ibid., 265.

6. Ibid., 274. Another philosopher working on this issue, both in general and specifically with respect to ecological restoration, is Martin Drenthen. See, for example, his "Reading Ourselves through the Land: Landscape Hermeneutics and Ethics of Place," in *Placing Nature on the Borders of Religion, Philosophy and Ethics,* ed. Forrest Clingerman and Mark H. Dixon (Farnham: Ashgate, 2011), 123–38.

7. Murray Bookchin has also suggested the need for such a project, though with a very different orientation. See his "Toward a Philosophy of Nature—The Bases for an Ecological Ethics," in *Deep Ecology*, ed. Michael Tobias (San Diego: Avant Books, 1985), 213–39.

8. I describe her view in much greater detail in "Developing Val Plumwood's Dialogical Ethical Ontology and Its Consequences for a Place-Based Ethic," *Ethics and the Environment* 14, no. 2 (2009): 39–55.

9. Val Plumwood, *Feminism and the Mastery of Nature* (New York: Routledge, 1993), 48–55 (hereafter cited as *FMN*).

10. Chapter 5 of *Feminism and the Mastery of Nature* argues this point at length. Charles S. Brown also argues that Cartesian dualism becomes, in the contemporary philosophical landscape, a Cartesian monism rather than an authentic naturalism. "The Real and the Good: Phenomenology and the Possibility of an Axiological Rationality," in *Eco-Phenomenology: Back to the Earth Itself*, ed. Charles S. Brown and Ted Toadvine (Albany: State University of New York Press, 2003), 7.

11. Neil Evernden, "The Fragile Division," in *The Social Creation of Nature* (Baltimore: Johns Hopkins University Press, 1992), 88–103 (hereafter cited as *SCN*).

12. The reason for his assertion lies in the historical character of the concept of nature, which he and others have also described in detail. Other accounts include Carolyn Merchant's more recent *Reinventing Eden: The Fate of Nature in Western Culture* (New York: Routledge, 2004); Pierre Hadot's *The Veil of Isis: An Essay on the Idea of the History of Nature*, trans. Michael Chase (Cambridge: Belknap, 2006); and Max Oelschlaeger's *The Idea of Wilderness: from Prehistory to the Age of Ecology* (New Haven, CT: Yale University Press, 1993).

13. Arne Naess, *Ecology, Community and Lifestyle: Outline of an Ecosophy*, trans. and ed. David Rothenberg (Cambridge: Cambridge University Press, 1993), 67.

14. Val Plumwood, "Toward a Progressive Naturalism," in *Recognizing the Autonomy of Nature: Theory and Practice*, ed. Thomas Heyd (New York: Columbia University Press, 2005), 44.

15. Ibid., 48.

16. Ibid., 47.

17. Steven Vogel, "Environmental Philosophy after the End of Nature," *Environmental Ethics* 24 (2002): 26–27.

18. Tom Regan, "The Nature and Possibility of an Environmental Ethic," *Environmental Ethics* 3 (1981): 20. The other major statement of the need for an environmental ethic that is distinct from a humanistic ethic extended to include duties to nonhuman nature is Richard Routley's "Is There a Need for a New, an Environmental, Ethic?" *Proceedings, 15th World Congress of Philosophy* 1 (1973): 205–10.

19. This point, of course, is also a disputed one. Norton, for example, objects to the fact that his own "weak anthropocentric" perspective does not count as an environmental ethic, but does little to argue against Regan's point that any anthropocentric ethic is, at best, a management ethic. Rather, he focuses on Regan's later positive arguments dealing with the supposed need to establish the "inherent goodness" of nature. Norton goes on to offer an explicitly managerial ethic, even using the example of a fiduciary trust to depict humanity's relation to nature, which only serves to buttress Regan's original claim. See Bryan G. Norton, "Environmental Ethics and Weak Anthropocentrism," *Environmental Ethics* 6 (1984): 131–48.

20. Deep ecologists, for example, claim to be basing their normative claims on cosmological rather than moral bases but still accept both of Regan's conditions. See, for example, Warwick Fox, *Toward a Transpersonal Ecology: Developing New Foundations for Environmentalism* (Albany: State University of New York Press, 1995), esp. chap. 6; or Naess's *Ecology, Community and Lifestyle*, esp. chap. 2.

21. Aldo Leopold, *A Sand County Almanac* (New York: Ballantine, 1970), 240 (hereafter cited as *SA*). I do want to call attention to the distinction here between being "antihumanistic" and being "nonanthropocentric" due to the debate between Norton and J. Baird Callicott regarding whether Leopold was an anthropocentrist. In a sense, Norton's case is that one can be antihumanistic (i.e., not believe that humans are somehow special or distinct from other natural beings) and still be an anthropocentrist (i.e., prioritize human needs and values in choosing how to act). Regardless of whether he is correct, both he and Callicott agree about Leopold's antihumanism in the sense I am using it.

Tim Hayward makes a case for an antihumanistic anthropocentrism being the strongest form of environmentalism. In many ways like Norton, he believes it eliminates the elements of speciesism and human chauvinism that other environmentalists have found so problematic, while still retaining a normative focus on human interests that is useful in the policy sphere. See Hayward, "Anthropocentrism: A Misunderstood Problem," *Environmental Values* 6, no. 1 (1997): 49–63.

22. Murray Bookchin, "Social Ecology Versus 'Deep' Ecology: A Challenge to the Ecology Movement," *Green Perspectives: Newsletter of the Green Program Project* 45 (1987): 1–23.

23. Vogel, "Environmental Philosophy," 28.

24. Steven Vogel, *Against Nature: The Concept of Nature in Critical Theory* (Albany: State University of New York Press, 1996), 6–7 (hereafter cited as *AN*).

25. Vogel discusses sympathetically how to include nonhuman animals in discourse ethics (*AN*, 151–60), but their fates remain tied to human discourse for the reason that all value comes from "the possibility of discursive redemption of a validity claim, and hence . . . those language-using entities who can engage in discourse" (161).

26. One of the most succinct expressions of this view can be found in Lynn White Jr.'s "The Historical Roots of Our Ecologic Crisis," *Science* 155 (1967): 1203–7.

27. I intend the phenomenological tradition to be understood quite broadly to include strands of French thought that are not technically phenomenological, for example, those that formulate themselves explicitly as responses to the Husserlian school (e.g., Derrida, Deleuze, and Foucault). Even though those figures would never call themselves phenomenologists, their way of thinking emerged from within phenomenology and in many ways shares the same concerns as it. Of course, there will be no one conception of nature operant throughout such an inclusive conception of the tradition, but I hope to draw insights from beyond what is conventionally recognized as phenomenological philosophy in an attempt both not to ignore the objections that have been made to phenomenological philosophy and to incorporate the insights of those criticisms into our understanding of how phenomenology must be done.

28. It would, of course, take us rather far afield to defend this claim here, but those familiar with the tradition will at least acknowledge the extent to which phenomenologists have explicitly thematized nature in a way that falls prey neither to the traditional mechanistic nor to the teleological understandings thereof. Consider, for example, Husserl's attempts in *Ideas II* to define nature as a phenomenon emerging from the personalistic attitude rather than as the foundation for it. Edmund Husserl, *Ideas Pertaining to a Pure Phenomenology and to a Phenomenological Philosophy*, second book: *Studies in the Phenomenology of Constitution*, trans. Richard Rojcewicz and André Schuwer (Dordrecht: Kluwer, 1989).

29. I have in mind here Michael Zimmerman's works "Beyond Humanism: Heidegger's Understanding of Technology," *Listening* 12 (1977): 74–83; "Toward a Heideggerian *Ethos* for Radical Environmentalism," *Environmental Ethics* 5 (1983): 99–131; and "Implications of Heidegger's Thought for Deep Ecology," *Modern Schoolman* 64 (1986): 19–43; as well as Bruce V. Foltz's *Inhabiting the Earth: Heidegger, Environmental Ethics, and the Metaphysics of Nature* (Atlantic Highlands, NJ: Humanities Press, 1995) in particular. Though Zimmerman recants much of his earlier analyses due to Heidegger's politics, they are still worth considering. For more details surrounding his "conversion," see his *Contesting Earth's Future: Radical Ecology and Postmodernity* (Berkeley: University of California Press, 1997), esp. 4–6.

30. "One sometime encounters what may be called situations or states of domination in which the power relations, instead of being mobile, allowing the various participants to adopt strategies modifying them, remain blocked, frozen." Michel

Foucault, "The Ethics of the Concern for the Self as a Practice of Freedom," in *Ethics: Subjectivity and Truth*, ed. Paul Rabinow (New York: New Press, 2006), 283.

31. Ted Toadvine, *Merleau-Ponty's Philosophy of Nature* (Evanston, IL: Northwestern University Press, 2009), 16.

CHAPTER 1: THE PROMISE OF A COMMON WORLD

1. Bruno Latour, *Politics of Nature*, trans. Catherine Porter (Cambridge, MA: Harvard University Press, 2004), 9 (hereafter cited as *PN*). Also, Latour uses italics liberally in his writings, so unless otherwise noted, the emphasis in quotations is his own.

2. Bruno Latour, *We Have Never Been Modern*, trans. Catherine Porter (Cambridge, MA: Harvard University Press, 1993), 9 (hereafter cited as *NBM*).

3. See, for example, *NBM*, 12; *PN*, 3; or his *Pandora's Hope: Essays on the Reality of Science Studies* (Cambridge, MA: Harvard University Press, 1999), 295 (hereafter cited as *PH*).

4. The list Latour supplies here includes "frozen embryos, expert systems, digital machines, sensor-equipped robots, hybrid corn, data banks, psychotropic drugs, whales outfitted with radar sounding devices, gene synthesizers, [and] audience analyzers" (NBM, 49–50). The list, obviously, could be iterated indefinitely, but what is of note is that its members range from phenomena that could be regarded as wholly natural but are discovered only through technological means (e.g., microbes), to ones that could be regarded as wholly cultural but rely upon natural processes for their agency (e.g., pharmaceuticals).

5. I am inclined to agree with Graham Harman that the reason for Latour's shifting away from the language of hybrids in his more recent work lies in the fact that "hybrid" retains the sense of "a mixture of pure ingredients," something with which Latour concerns himself in *We Have Never Been Modern*. Support for this view can also be found at *NBM*, 77–79. The question of "mixture" will be revisited in chapter 3. Graham Harman, *Prince of Networks: Bruno Latour and Metaphysics* (Melbourne: re.press, 2009), 63.

6. Val Plumwood's book *Environmental Culture: The Ecological Crisis of Reason* (New York: Routledge, 2002), deals with this theme in depth. Take, for example, her claim that "reading embodied action is part of all our lives, and is the common language of embodied beings" (192). David Abram's work in *The Spell of the Sensuous: Perception and Language in a More-Than-Human World* (New York: Vintage, 1996), begins from the same notion that bodies are inherently meaningful and will be considered directly in chapter 3.

7. Plumwood's *Environmental Culture* stands as a rigorous and sensitive treatment of these subjects.

8. In his own words, what he seeks is "to retain the production of a nature and of a society that allow changes in size through the creation of an external truth and a subject of law, but without neglecting the co-production of sciences and societies" (*NBM*, 134).

9. Though the organismic conception of nature is not common to *all* views that oppose mechanism, it is one of the most prevalent. Even Plumwood succumbs to this temptation, arguing that nature should be attributed "mind-like qualities" in a teleological vein. I criticize this idea in greater detail in "Developing Val Plumwood's

Dialogical Ethical Ontology and Its Consequences for a Place-Based Ethic," *Ethics and the Environment* 14, no. 2 (2009): 39–55.

10. I recognize that the situation is a complex one given statements of Latour's such as the following: "Those who speak of 'disciplining' and of 'domination' can begin their analysis only at the point where almost everything is over, where power has no longer to be 'negotiated,' where all that remains is to convince those at the bottom of the ladder." Latour, *The Pasteurization of France*, trans. Alan Sheridan and John Law (Cambridge, MA: Harvard University Press, 1988), 140. My understanding of the situation, as I mentioned in the previous section, is that Latour would agree that nonhumans in general are subject to domination, but that the domination also recoils on humanity, dominating us in turn. The situation also works in the reverse: the more the nonhuman actants press in upon us, the more our agency modifies them (*PH*, 281). Quotes such as the one above are oriented toward criticizing the focus on domination in a traditionally political sense (i.e., the overt exercise of power upon the oppressed), while ignoring the subtler ways in which that power is institutionalized and mobilized through the cooperation and resistance of nonhuman agency as well.

11. As we will discuss in chapter 3, Merleau-Ponty will use the Russian revolution to reinforce this point: communism merely inscribed a different form of oppression through the establishment of a different conception of transcendent nature.

12. I qualify the sense in which modernity considers nature to be unchanging because, of course, the Darwinist and ecological conceptions of nature admit to a great amount of change within nature. In both cases, however, the underlying conception of nature is that of a transcendent, biological order of some sort. Saying this is not meant to be a challenge to the validity of Darwinism or ecology, but rather simply to indicate that the level at which nature is understood as unchanging is not the phenomenal but rather that of the laws that make the mechanisms that underlie that world.

13. See chapters 3 and 4 of *Pandora's Hope* for a more detailed account of the institution of microbes as a substance.

14. He elaborates this in greater detail in *Politics of Nature* (42–49) and *We Have Never Been Modern* (91–129), but many of his actual studies are conceived as an anthropological study of the practices of modernity. See, for example, the introduction to "Circulating Reference" (*PH*, 24); or his comments in *We Have Never Been Modern* (7, 15) concerning the need for anthropologists to study themselves as modernists.

15. The term is intended more as an opposition to the universe of classical physics than to refer to the many-worlds interpretation of quantum theory.

16. Gilles Deleuze and Félix Guattari, *A Thousand Plateaus: Capitalism and Schizophrenia*, trans. Brian Massumi (Minneapolis: University of Minnesota Press, 1987), esp. chap. 6.

17. Ordinarily, this would be the place to recount Latour's views about "the sciences" as opposed to "Science," but in many ways that story has already been told in what has come before. To overgeneralize what is really a rich and nuanced view, Science is a modern conception, and one that requires a monolithic nature that it is to discover through its objective methods. The sciences behave exactly as Science does, but they acknowledge (1) the sociality of scientific practices (e.g., colloquia, laboratories, journals, experimental technicians, etc.); and (2) that the truths developed through scientific practice are historical articulations of the world and not its absolute ahistorical structure. These two acknowledgments change dramatically little

about what science does, but change rather significantly the meaning of science and its role in society.

18. As we saw with the antimoderns, these values are not necessarily going to be the ones valued by the moderns, and so there are some deep resonances between this project and Nietzsche's call to revaluate all the highest values.

19. The situation is exacerbated both by Latour using the language of "entelechies" rather freely in "Irreductions," the second part of *The Pasteurization of France*, and how Schelling uses the term "actant" in an overtly teleological manner. See F. W. J. Schelling, *First Outline of a System of the Philosophy of Nature,* trans. Keith R. Peterson (Albany: State University of New York Press, 2004), esp. 23–24, where he defines *actants* in terms of "natural monads."

20. With typical Latourian wit: "Nothing proves, however, that the assembly is going to come off well, that the participants are all going to find themselves in the ecumenical equivalent of some Woodstock festival in honor of Gaia. We still do not have the slightest idea what the consequences of such a reunion, such a resumption of the work of collection, might be" (*PN*, 82).

21. "I never *act*; I am always slightly surprised by what I do. That which acts through me is also surprised by what I do, by the chance to mutate, to change, and to bifurcate, the chance that I and the circumstances surrounding me offer to that which has been invited, recovered, welcomed" (*PH*, 281).

22. Jane Bennett, *Vibrant Matter: A Political Ecology of Things* (Durham, NC: Duke University Press, 2010), esp. chap. 1 and 4. On the point of the autonomy of nature, the essays collected in Thomas Heyd's anthology, *Recognizing the Autonomy of Nature: Theory and Practice* (New York: Columbia University Press, 2005), are exemplary.

23. This is one of the many reasons why I do not follow Harman in arguing for an "object-oriented philosophy" in the way he describes. To say that "objects exist as autonomous units, but they also exist in conjunction with their qualities, accidents, relations, and moments without being reducible to these," obscures the point that objects are events (Harman, *Prince of Networks*, 156). Of course, Harman recognizes this and faults Latour for that belief (156). While I hope the view of nature I develop will be "object-oriented" in a broad sense, I will argue that an individual object is, as Latour suggests, a nexus of relations that lacks the kind of autonomy Harman attributes to them: there is no thing independent of its properties; things simply are the totality of their potential relations.

CHAPTER 2: SCIENCE, TECHNOLOGY, AND THE CLOSURE OF NATURE

1. In "Latour's Heidegger" (*Social Studies of Science* 40, no. 4 [2010]: 579–98), Jeff Kochan, for example, argues that Latour engages in "a tactical dissimulation meant to direct attention away from the weaknesses in his own account" (580).

2. Søren Riis, "The Symmetry Between Bruno Latour and Martin Heidegger: The Technique of Turning a Police Officer into a Speed Bump," *Social Studies of Science* 38, no. 2 (2008): 285–301. He argues that Latour's vision of the world fits neatly into the model of "enframing" Heidegger describes. In other words, he believes Latour's philosophy results in a situation where "all traits of individuality are erased, and all humans have turned into a collective mass of computer-humans and other hybrid artefacts. Humans are only defined as a temporary constellation of properties, to which

'purposeful actions' and 'intentionality' do not belong. All human properties can in principle be transferred to non-humans—there is nothing essential to being human" (295). As we have already seen in the last chapter, the collective does no such thing since it refers not to an undifferentiated amalgam of bodies, but to concrete actants whose differences are critical to understanding how to create the best possible common world. Though there are definite similarities in their accounts, then, I will argue that they do not move in this direction.

3. It is not my intention to delve into an in-depth analysis of the congruence between their philosophies in general, both because it would distract from the central theme of this inquiry and because others have already commented on the matter competently. Even though there is dispute concerning the relationship, I do not believe there is need for the current study to take up a position in the general debates on the subject. For a small sampling in addition to the already cited essays by Kochan and Riis, please see Lynnette Khong, "Actants and Enframing: Heidegger and Latour on Technology," *Studies in the History and Philosophy of Science* 34 (2003): 693–704; Kasper Schiølin, "Follow the Verbs! A Contribution to the Study of the Heidegger-Latour Connection," *Social Studies of Science* 42, no. 5 (2012): 775–86; and Graham Harman, "The Importance of Bruno Latour for Philosophy," *Cultural Studies Review* 13, no. 1 (2007): 31–49.

4. This point regarding nature as a system will recur throughout the chapter. It is also the point Kochan utilizes to criticize Latour's views as distinctively modern (see "Latour's Heidegger," 585–86). As I will argue in the text, when Heidegger is speaking of nature as a system, he is referring more to the idea that there is an inherent order to the surrounding world, that beings can be ordered and organized in some systematic way independent of a projecting open of being, by which he means a method of understanding the phenomena one encounters as a whole.

5. I have in mind here Bruce V. Foltz and Thomas Sheehan in specific. Foltz argues that *physis* is best understood for Heidegger as "primordial nature." Foltz, *Inhabiting the Earth: Heidegger, Environmental Ethics, and the Metaphysics of Nature* (Atlantic Highlands, NJ: Humanities Press, 1995), 80. For reasons I will argue in the chapter itself, I find this to be highly problematic. There is much of value in his extremely thoughtful, careful, and exacting account of Heidegger's philosophy, and much of what is said below concerning nature accords with what is explained in Foltz's text. If, however, the analysis of nature's place in the history of being is accurate, then it is going to be necessary to dissociate ourselves from the traditional conception of it, including the Greek one, in order to "begin again." The account that I will offer in the next two chapters, then, will differ rather significantly from Foltz's nature, even as I continue to find value in the idea of poetic dwelling as he describes it. Poetic dwelling retains its importance because, as we shall see, what Heidegger seeks is a new mode of relation and interacting with the beings around us, not a practice of isolation and withdrawal, as is the tendency in a great deal of environmental philosophy. Sheehan's views will be more closely discussed in what follows, so I will momentarily postpone comment.

6. Quentin Meillassoux, *After Finitude: An Essay on the Necessity of Contingency*, trans. Ray Brassier (New York: Continuum, 2010). I withhold specific comment on his views for separate treatment elsewhere, though much of what I formulate in the coming pages can serve as a way of responding to some of Meillassoux's blanket dismissals of all forms of phenomenology.

7. A basic exposition of the parenthesizing of the natural attitude can be found in Edmund Husserl, *Ideas Pertaining to a Pure Phenomenology and to a Phenomenological Philosophy,* first book: *General Introduction to a Pure Phenomenology*, trans. F. Kersten (The Hague: Martinus Nijhoff, 1983), 57–62.

8. This naturalization usually takes one of two forms, either reducing intentional relationships to relations between objects, as do philosophers like J. Kevin O'Regan, Alva Noë, and Francisco J. Varela; or the more promising route of showing the necessary contribution of objects to their own appearance to consciousness, as does David Wood. See O'Regan and Noë, "A Sensorimotor Account of Vision and Visual Consciousness," *Behavioral and Brain Sciences* 24, no. 5 (2001): 939–73; Noë, *Action in Perception* (Cambridge, MA: MIT Press, 2004); Varela, "The Specious Present: A Neurophenomenology of Time Consciousness," in *Naturalizing Phenomenology: Issues in Contemporary Phenomenology and Cognitive Science*, ed. Jean Petitot et al. (Stanford, CA: Stanford University Press, 1999), 266–306; and Wood, "What Is Eco-Phenomenology?" in *Eco-Phenomenology: Back to the Earth Itself*, ed. Charles S. Brown and Ted Toadvine (Albany: State University of New York Press, 2003), 211–33. Scott L. Marratto offers a thoroughgoing analysis of the tendency to naturalize subjectivity in his recent *The Intercorporeal Self: Merleau-Ponty on Subjectivity* (Albany: State University of New York Press, 2012).

9. There is a third on her account as well, "The key to understanding technology does not lie in resolving the debate between technological determinism and social determinism," but delving into this point would be to digress significantly. Khong, "Actants and Enframing," 695–98.

10. I am, of course, claiming neither that *all* philosophy must be immediately practical, nor that such philosophers do not already exist. Irene J. Klaver's phenomenologically informed works on water, boundaries, and stone are commendable in this regard. See, for example, "Boundaries on the Edge," in *Nature's Edge: Boundary Explorations in Ecological Theory and Practice*, ed. Charles S. Brown and Ted Toadvine (Albany: State University of New York Press, 2006), 146–76.

11. Martin Heidegger, *Gesamtausgabe*, vol. 65, *Beiträge zur Philosophie* (Frankfurt: Klostermann, 1989), 48. Translated by Parvis Emad and Kenneth Maly as *Contributions to Philosophy (from Enowning)* (Bloomington: Indiana University Press, 1999), 34. Hereafter cited as *BP*, with the original language pagination preceding that of the translation.

12. Martin Heidegger, *Gesamtausgabe*, vol. 40, *Einführung in die Metaphysik* (Frankfurt: Klostermann, 1983), 15–19. Translated by Gregory Fried and Richard Polt as *Introduction to Metaphysics* (New Haven, CT: Yale University Press, 2000), 14–18. Hereafter cited as *EM*, with the original language pagination preceding that of the translation.

13. Martin Heidegger, "Vom Wesen und Begriff der Physis Aristoteles, *Physik* B, 1," in *Gesamtausgabe,* vol. 4, *Wegmarken* (Frankfurt: Klostermann, 1967), 237. Translated by Thomas Sheehan as "On the Essence and Concept of *Physis* in Aristotle's *Physics* B, I," in *Pathmarks*, ed. William McNeill (Cambridge: Cambridge University Press, 1998), 183. Hereafter cited as WBP, with the original language pagination preceding that of the translation.

14. For a small but diverse selection, see the following: Eileen Crist, "Against the Social Construction of Nature and Wilderness," *Environmental Ethics* 26 (2004): 5–24; David W. Kidner, "Fabricating Nature: A Critique of the Social Construction

of Nature," *Environmental Ethics* 22 (2000): 339–57; George Sessions, "Postmodernism and Environmental Justice: The Deminse of the Ecology Movement?" *Trumpeter* 12, no. 3 (1995): 150–54; Sessions, "Reinventing Nature . . . ? A Response to William Cronon's *Uncommon Ground*," *Trumpeter* 13, no. 1 (1996): 33–38; and Michael E. Soulé, "The Social Siege of Nature," in *Reinventing Nature? Responses to Postmodern Deconstruction*, ed. Michael E. Soulé and Gary Lease (Washington, DC: Island Press, 1995), 137–70. In fact, many of the essays in that last volume would suffice.

15. One of Heidegger's clearest descriptions of this transition, which we will treat in greater detail below, comes in "Die Frage nach der Technik," in *Gesamtausgabe*, vol. 7, *Vorträge und Aufsätze* (Frankfurt: Klostermann, 2000), 7–36. Translated by William Lovitt as "The Question Concerning Technology," in *The Question Concerning Technology and Other Essays* (San Francisco: Harper Colophon, 1977), 3–35. Hereafter cited as FT, with the original language pagination preceding that of the translation.

16. Heidegger uses the German word *Wesen*, which would conventionally translate as "essence," "nature," or "being," as a verb rather than as a noun. This "verbal *Wesen*," as Merleau-Ponty will later say, is extremely difficult to translate insofar as there is no word in English that captures the relation that Heidegger is bringing to the fore by making the substantive into a verb. Some translators use "holds sway" for *west*, but *Walten* is also translated as "sway" or "prevails"; others use "presences" because of the separability of the "*an*" in "*anwesen*," but this can also cause confusion due to Heidegger's attempts to distance the notion of essence from that of presence alone. I hope to avoid any confusion by remaining literal and using "essences" for *west*. Doing so, of course, is rather awkward, but I hope that it will make more transparent the transition in sense in the notion of essence from the traditional sense of "what makes a being what it is" to Heidegger's own of how a being simultaneously manifests and conceals itself.

17. See also section 11 of Martin Heidegger, *Gesamtausgabe,* vol. 6, *Nietzsche I* (Frankfurt: Klostermann, 1996). Translated by David F. Krell as *Nietzsche,* vol. 1–2 (San Francisco: HarperCollins, 1991).

18. For example, "World itself is not an innerworldly being, and yet it determines innerworldly beings to such an extent that they can only be encountered and discovered and show themselves in their being because 'there is' world." Martin Heidegger, *Sein und Zeit* (Tübingen: Niemeyer, 2001), 72. Translated by Joan Stambaugh as *Being and Time* (Albany: State University of New York Press, 1996), 67–68. Hereafter cited as *SZ*, with the original language pagination preceding that of the translation.

19. A conventional translation of fragment 53 reads, "War is the father of all and king of all, and has shown some as gods and others as human beings, made some slaves and others free."

20. For a much more detailed, nuanced, and rich discussion of precisely how this occurs, see Martin Heidegger, "Der Ursprung des Kunstwerkes," in *Holzwege* (Frankfurt: Klostermann, 1957), 7–68. Translated by Julian Young and Kenneth Haynes as "The Origin of the Work of Art," in *Off the Beaten Track* (New York: Cambridge University Press, 2002), 1–56. Hereafter cited as UK, with the original language pagination preceding that of the translation.

21. In *Metaphysics*, Z III, Aristotle says that *ousia* is firstly *hypokeimenon*, which in turn can be meant as either *morphe, hyle*, or a combination of the two (*synolon*), a concrete thing (1029a2–5). See Aristotle, *The Metaphysics*, bks. 1–9, trans. Hugh Tredennick (Cambridge, MA: Harvard University Press, 1956).

22. Heidegger indirectly asserts this in a different context, within which he rejects the characterization of being as either finite or infinite because being, which is not itself a being, can have no magnitude. See *BP*, 268/189.

23. A more conventional translation of these lines is, "It is, then, the form that is nature; but since 'form' and 'nature' are ambiguous terms, inasmuch as shortage is a kind of form . . ." (193b19–21). Aristotle, *The Physics*, vol. 1, bks. 1–4, trans. Philip H. Wicksteed and Francis M. Cornford (Cambridge, MA: Harvard University Press, 1957).

24. The following statement is emblematic of this point: "The absence of a sign can be a sign, and expression is not the adjustment of an element of discourse to each element of sense, but an operation of language upon language that suddenly decenters itself toward its sense." Maurice Merleau-Ponty, "Le langage indirect et les voix du silence," in *Signes*, folio ed. (Paris: Gallimard, 1960), 71. Translated by Michael B. Smith and revised by Leonard Lawlor as "Indirect Language and the Voices of Silence," in *The Merleau-Ponty Reader*, ed. Ted Toadvine and Leonard Lawlor (Evanston, IL: Northwestern University Press, 2007), 245. Hereafter cited as LI, with original language pagination preceding that of the translation.

25. See Aristotle's *Metaphysics,* Z VIII, within which he describes in detail the relation between *morphe* and *hyle* within a being in terms of generation.

26. Aristotle, *Metaphysics*, 1033b18–1034a1.

27. These are not, of course, the only discussions of this fragment. Others, for example, appear in "On the Question of Being" and in *The Principle of Reason*. The sense of all these discussions is similar, and so I limit myself to those that appear in the texts already discussed.

28. The idea contained in the conventional translation of this fragment, that if we could merely see what nature is hiding then we would know nature itself, has been the launching point for scientific endeavors for centuries. Pierre Hadot offers a more in-depth consideration of the genealogy of this fragment and its impact on the Western scientific mind in *The Veil of Isis: An Essay on the History of the Idea of Nature*, trans. Michael Chase (Cambridge, MA: Belknap, 2006).

29. Bernard Williams has a much more detailed and critical engagement with Aristotle's views on this subject in *Shame and Necessity* (Berkeley: University of California Press, 1993), esp. chap. 5.

30. Roderick Nash provides a historical example of this conflict between resource and aesthetic values with respect to the failure of plans to dam the Colorado River as it passes through the Grand Canyon. See Nash, *Wilderness and the American Mind*, 4th ed. (New Haven, CT: Yale University Press, 2001), 227–37.

31. The rhetoric at present, of course, is slightly different from the one proposed above, revolving primarily around the idea of "energy independence." I would argue, however, that it is the notion of the land as a use-object that makes the project of energy independence possible. Energy independence is predicated upon the belief that there are vast, untapped deposits of energy within America that, if we could simply challenge the land a bit further, would supply the demand for energy.

32. For more detailed expositions of Heidegger's views on the sciences, see Trish Glazebrook, *Heidegger's Philosophy of Science* (New York: Fordham University Press, 2000). Though this text proposes a reinterpretation of Heidegger's works as a whole through his understanding of science, a task I am not convinced is always a constructive one, it does lay out in a systematic manner the central ideas of Heidegger's views on science. Don Ihde presents an account of Heidegger's views on science and

technology more consonant with my goals in this text in *Heidegger's Technologies: Postphenomenological Perspectives* (New York: Fordham University Press, 2010). Even though Ihde is critical of Heidegger's views throughout, the direction in which he is pushing the philosophy is, I believe, a productive one.

33. As an essential project of *Dasein*, science remains within the structure of being-in-the-world: "In knowing, Dasein gains a new *perspective of being* toward the world always already discovered in Dasein. This new possibility of being can be independently developed. It can become a task, and as scientific knowledge can take over the guidance for being-in-the-world. . . . Knowing is a mode of Dasein which is founded in being-in-the-world. Thus, being-in-the-world, as a fundamental constitution, requires a *prior* interpretation" (*SZ*, 62/58). Science cannot, then, be used to explain being-in-the-world for the simple reason that it takes place only upon its basis.

34. For a detailed example of how this works in terms of the biological sciences, see the second lecture course in Martin Heidegger, *Gesamtausgabe*, vol. 29–30, *Die Grundbegriffe der Metaphysik: Welt—Endlichkeit—Einsamkeit* (Frankfurt: Klostermann, 1983). Translated by William McNeill and Nicholas Walker as *The Fundamental Concepts of Metaphysics: World, Finitude, Solitude* (Indianapolis: Indiana University Press, 1995).

35. David E. Cooper suggests in his "Heidegger on Nature" (*Environmental Values* 14 [2005]: 339–51) that Heidegger's so-called environmental ethic depends upon his critique of science, but this claim seems a bit overstated given that the path that opened up the possibility for machination, which is predominantly responsible for the current state of the environment for Heidegger, existed long before the advent of modern science. In other words, Cooper is right concerning the centrality of Heidegger's views on science for extending Heidegger's thinking to the environment, but the emphasis should be placed on how "scientific realism" and the "purely pragmatic" use of nature "hang together," a fact Cooper himself points out (cf. esp. 346–48), and not in making the critique of science foundational. As with all things Heidegger, the questioning of being, its truth, and history will always be at the center of his analysis.

36. This has recently been a much commented upon aspect of Heidegger's thought. Heidegger explains his own views on the subject in *The Fundamental Concepts of Metaphysics*. For a sympathetic interpretation of Heidegger, see William McNeill, "Life Beyond the Organism: Animal Being in Heidegger's Freiburg Lectures, 1929–30," in *Animal Others: On Ethics, Ontology, and Animal Life*, ed. H. Peter Steeves (Albany: State University of New York Press, 1999), 197–248; or chapter 3 of Brett Buchanan, *Onto-Ethologies: The Animal Environments of Uexküll, Heidegger, Merleau-Ponty, and Deleuze* (Albany: State University of New York Press, 2008).

37. Riis, "Symmetry Between Bruno Latour and Martin Heidegger," 297.

38. Kochan, "Latour's Heidegger," 586; Riis, "Symmetry Between Bruno Latour and Martin Heidegger," 295.

39. Kochan, "Latour's Heidegger," 584.

CHAPTER 3: THE OPENING OF THE EARTH

1. For example, Thomas Sheehan asserts that Heidegger's "program" involves a continual philosophical "retreat from *techne* toward *physis*." See Sheehan, "Nihilism: Heidegger/Jünger/Aristotle," in *Phenomenology: Japanese and American Perspectives*, ed. Burt C. Hopkins (Dordrecht: Kluwer, 1998), 299. Though exemplary in its

explanation of the movements of Heidegger's thought with respect to Aristotle, Sheehan's article overlooks certain elements of Heidegger's thinking in making his claim that there is no escape from nihilism. We will return to this claim since it is the nihilistic essence of metaphysics that must be displaced in order to open up another thought of nature.

2. Ihde argues the point that Heidegger's views require deromanticizing in chapter 3 of *Heidegger's Technologies: Postphenomenological Perspectives* (New York: Fordham University Press, 2010). Ihde somewhat misses Heidegger's point when he criticizes him for a "nostalgic merging of art and technology" (76). Heidegger does not desire some return to an original state that has been lost or have a nostalgic preference for older technologies over the newer. His point is that technology and art even today remain grounded upon a specific interpretation of being. While Ihde is still correct that the distinction can at times *seem* to be based in nostalgia, it is important also to note that what is being evaluated is the *type* of gathering the technology embodies, whether the technology does or does not embody the limitation of possible interpretations of being present within machination. Of course, Latour may be correct here that any technology can meet this requirement if we are more charitable in our understanding of actants, but I will remain uncommitted with respect to that question.

3. Jeff Kochan, "Latour's Heidegger," *Social Studies of Science* 40, no. 4 (2010): 584.

4. See Michael E. Zimmerman, "Heidegger's Phenomenology and Contemporary Environmentalism," in *Eco-Phenomenology: Back to the Earth Itself*, ed. Charles S. Brown and Ted Toadvine (Albany: State University of New York Press, 2003), 73–101. In this essay Zimmerman takes up Sheehan's claims with respect to nihilism in order to argue that though Heidegger speaks persuasively about the need to address the devastation of the earth, his philosophy does not offer sufficient resources to actually assuage the damage. Since his latter claim hinges on Sheehan's arguments, we will devote our attention to those arguments.

5. Leibniz, in a letter to Arnauld (April 30, 1687), claims that "being is one thing and beings are another; but the plural presupposes the singular, and where there is no being still less will there be several beings. What could be clearer?" See G. W. F. Leibniz, *Philosophical Essays*, trans. Roger Ariew and Daniel Garber (Indianapolis, IN: Hackett, 1989), 86. F. W. J. Schelling makes a distinction between "being itself" and that which is "only a singular form or kind of being" in *First Outline of a System of the Philosophy of Nature*, trans. Keith R. Peterson (Albany: State University of New York Press, 2004), 13.

6. This contention is of particular importance in responding to Sheehan's and Zimmerman's claim that Heidegger can offer no nondominating opposition to the worldview of machination. Sheehan's accurate claim that all worldviews involve mastery can remain intact while at the same time recognizing that Heidegger is not advocating a worldview to replace the current one. His actual aim is to displace the idea of a worldview entirely. While it is true that thinking in terms of the representation of beings entails "the availability of absolutely everything to human cognition and will" (Sheehan, "Nihilism," 295), it is for precisely this reason that Heidegger's ultimate "ethical" mandate is "*Gelassenheit*," which I will later describe as behaviors that allow interaction with beings without challenging them, in Heidegger's sense of the word, through representation. These behaviors, however, can only arise from the recognition of how the projecting open of being might dominate beings and remaining mindful of this in our interactions with beings.

7. This point is essential to understanding why Heidegger is not arguing for a cultural or subjective relativism. Conflicts of this sort always take place *within a given understanding of the being of beings*. The change for which Heidegger calls would also require an abandonment of the idea of "culture" as a subjective and therefore representational concept in favor of the notion of a "people." If two peoples were genuinely to articulate beings in a distinct manner, there would be greater obstacles to their interaction than those that arise from differences in value. In this way, Heidegger is again very close to Latour's views on nature-culture, though they express their respective views very differently.

8. I recognize that this is a horrible English rendering, but no more so than "essential swaying," the usual translation. To risk a preliminary interpretation of *Wesung*, "essencing" should be heard in the sense of how being unconceals itself and concurrently conceals itself or withdraws through that unconcealment. In the same way that "essence" is used as a verb, *Wesung* is the substantive of this verbal *wesen*.

9. This point is the heart of Sheehan's argument, and one with which I clearly agree. The difference between us lies in the import of this point, with him arguing that there is no "escape from *Herrschaft* [mastery]" because humans will always project open being (Sheehan, "Nihilism," 297), and my maintaining that the openness of the questioning is sufficient to turn away from mastery without ever definitively dispatching the danger that projecting open being poses.

10. See Martin Heidegger, "Die Zeit des Weltbildes," in *Holzwege* (Frankfurt: Klostermann, 1997). Translated by Julian Young and Kenneth Haynes as "The Age of the World Picture," in *Off the Beaten Track* (New York: Cambridge University Press, 2002), 57–85. Hereafter cited as ZW, with the original language pagination preceding that of the translation.

11. This is why Sheehan's claim with respect to the inevitability of nihilism is false: nihilism can be assailed by means of the elimination of representational thinking. Thinking in terms of a representing subject is *not* inevitable, hence Heidegger's attempts to redefine "what is called thinking." So, once representation is no longer the dominant manner of thinking and thus the basis for our interactions with beings, it may then be possible to deal with a being without dominating it.

12. Heidegger provides much more detailed analysis of Hegel's form-content distinction in *Phenomenology of Spirit* insofar as it pertains to the concept of "experience" and the constitution of "consciousness" in his "Hegels Begriff der Erfahrung," in *Holzwege*, 115–208. Translated by Julian Young and Kenneth Haynes as "Hegel's Concept of Experience," in *Off the Beaten Track*, 86–156.

13. The translation of this word is vexing, to say the least. In their introduction to *Contributions to Philosophy*, Parvis Emad and Kenneth Maly offer a succinct explanation of the difficulties surrounding the term's usual translations as "event," "appropriation," "event of appropriation," or "befitting" (*BP*, xx–xxi). Their proposed solution to these difficulties—translating the word as "enowning"—does little to clarify the hermeneutic situation of the word given that part of Heidegger's project is making the *ordinary* use of German *peculiar*, thus demonstrating the power of language to be expressive beyond its ordinary employment. Contrary to their claim that leaving the word untranslated is itself a translation, I will not venture to translate the word in this text, though I will offer an interpretation that may lead one to provide an everyday English word that might serve similarly as a translation.

14. This is also the reason why Heidegger's search for the "truth of being" does not require that he pursue a fixed determination of or name for being. Rather, we should hear "truth of being" as "that which is true in the light of a given projecting open of being itself." In this latter sense, truth would genuinely be *of* being.

15. The verb form of *Ereignis* is *ereigen*, which can be translated conventionally as "occurs." In this case, "occurs" would be a completely appropriate translation, but I have left the verb untranslated for the same reason that I have left *Ereignis* so.

16. I must emphasize, however, that "letting a being be" does not mean "leaving [it] pristine and untouched" (Sheehan, "Nihilism," 297). Beings, including human beings, are what they are through their relation to other beings, not in total isolation from other beings. Identifying "letting be" with a removal from all relations would be to affirm an ipseity that exists prior to relation: the metaphysical conception of essence.

17. This, of course, is exactly what Zimmerman and Sheehan hold to be the case in the aforementioned articles.

18. For a much richer explanation of this process, see section 72 of *Being and Time*.

19. These are extremely difficult sentences to render into English, and so I supply the German here as well: "Raum ist die berückende Ab-gründung des Unhalts. / Zeit ist die entrückende Ab-gründung der Sammlung. / Die Berückung ist abgründiger Umhalt der Sammlung. / Die Entrückung ist abgründiger Sammlung auf den Umhalt."

20. For a brief discussion of this issue, see Martin Heidegger, *Gesamtausgabe*, vol. 66, *Besinnung* (Frankfurt: Klostermann, 1997), 289–90. Translated by Parvis Emad and Thomas Kalary as *Mindfulness* (London: Continuum, 2006), 257–58. Here, Heidegger argues that the distinctions between possibility, actuality, and necessity stem from the preunderstanding of being as constant presence and that therefore questions of modality belong to metaphysics, not the thinking of the other beginning.

21. See *SCN*, chap. 1.

22. Incidentally, this would also account for the competing interpretations of nature within ecology: if the projecting open is historically contingent, then opposing or contradictory interpretations ought to be *expected* within a science that takes a fixed projection as its starting point.

23. Though I translate *Gesichtskreis* as "prospect," the sense lies closer to the older word "ken": a range of vision or a place from which to look upon something. It can also mean a horizon opened from a given perspective, which is implied by all those other possible senses.

CHAPTER 4: MERLEAU-PONTY AND NATURE AS THE COMMON WORLD

1. Maurice Merleau-Ponty, *La structure du comportement* (Paris: Quadridge/PUF, 2006), 1. Translated by Alden Fisher as *The Structure of Behavior* (Pittsburgh, PA: Duquesne University Press, 1963), 3. Hereafter cited as *SC*, with the original language pagination preceding that of the translation.

2. David Abram, *The Spell of the Sensuous: Perception and Language in a More-Than-Human World* (New York: Vintage, 1996). Hereafter cited as *SS*.

3. See David Abram, "Merleau-Ponty and the Voice of the Earth," *Environmental Ethics* 10 (1988): 101–20.

4. The course notes on nature were first published in French in 1994, two years prior to the publication of Abram's text, and not translated into English until 2003. I

imagine Abram's book was written and awaiting publication prior to the availability of this material in French because the notes are not cited in his book.

5. Though I will introduce this idea here as well, I provide a more detailed account of flesh and its place in Merleau-Ponty's philosophy in my "Flesh and Nature: Understanding Merleau-Ponty's Relational Ontology," *Research in Phenomenology* 41 (2011): 327–57.

6. Renaud Barbaras, "Merleau-Ponty et la nature," *Chiasmi International* 2 (2000): 47–61. The following questions are raised at the conclusion of his essay, page 60.

7. Maurice Merleau-Ponty, *Phénoménologie de la perception* (Paris: Gallimard, 1945). Translated by Colin Smith as *Phenomenology of Perception* (New York: Routledge, 1962). Hereafter cited as *PhP*, with the original language pagination preceding that of the translation.

8. Maurice Merleau-Ponty, *La nature: Notes, cours du Collège de France* (Paris: Seuil, 1994), 20. Translated by Robert Vallier as *Nature: Course Notes from the Collège de France* (Evanston, IL: Northwestern University Press, 2003), 4. Hereafter cited as *N*, with the original language pagination preceding that of the translation.

9. Barbaras, "Merleau-Ponty et la nature," 51.

10. Ted Toadvine, *Merleau-Ponty's Philosophy of Nature* (Evanston, IL: Northwestern University Press, 2009), 38.

11. In the psychological literature Merleau-Ponty references to make his argument, *apprentissage* is not necessarily an abstract cognitive function, but rather is a bodily learning that takes place through repetition of bodily gestures.

12. Merleau-Ponty's relationship to critical thought is rather nuanced. Consider the following passage, which I reproduce here rather than in the text because of its length:

> Speaking generally, it seems that we are rejoining the critical idea. Whatever external conditions may be—bodily, psychological, social—upon which the development of consciousness depends and even if it is only gradually constituted in history, the history itself out of which it comes is only a view which consciousness gives itself with regard to the acquired consciousness of self. A reversal of perspective is produced vis-à-vis adult consciousness: the historical becoming which prepared it was not *before* it, it is only *for* it; the time during which it progressed is no longer the time *of* its constitution, but a time which it constitutes; and the series of events is subordinated to its eternity. Such is the perpetual reply of critical thought to psychologism, sociologism, and historicism.
>
> This discussion of causal thinking has seemed valid to us and we have pursued it at all levels of behavior. It leads, as we just said, to the transcendental attitude. This is the first conclusion which we have to draw from the preceding chapters. (*SC*, 222/206)

So, even though the critical philosophy that Merleau-Ponty argues for "stands in a relation of simple homonymy" with certain transcendental philosophies, Merleau-Ponty describes his own view as follows: "Each setting into form [*mise en forme*] appears to us on the contrary to be an event in the world of ideas, the institution of a new dialectic, the opening of a new region of phenomena, and the establishment of a new constitutive layer that eliminates the preceding one as isolated moment, but conserves and integrates it" (*SC*, 224/108). The language here is profoundly Hegelian, and so it should be no surprise that he favorably cites Hegel's definition of nature as "the exterior of a concept" (*SC*, 227/210). Thus, to say that Merleau-Ponty at this early stage of his philosophy has made a definitive break with critical philosophy is somewhat exaggerated. While true that he wants to push critical philosophy away from its intellectualist interpretations and toward a more realistic or naturalistic understanding of itself, it nevertheless remains the case that the forms through which we interact with the world are those of perceptual consciousness.

13. This last inquiry, of course, is the one that led James Lovelock to first formulate Gaia theory, within which the earth itself is taken to be a single, self-regulating system, that is, an organism. See his two books *Gaia: A New Look at Life on Earth* (New York: Oxford University Press, 2000); and *The Revenge of Gaia: Earth's Climate Crisis and the Fate of Humanity* (New York: Basic Books, 2007).

14. Toadvine, *Merleau-Ponty's Philosophy of Nature*, 21.

15. Maurice Merleau-Ponty, *La prose du monde*, ed. Claude Lefort (Paris: Gallimard, 1969). Translated by John O'Neill as *The Prose of the World* (Evanston, IL: Northwestern University Press, 1973). Hereafter cited as *PM*, with the original language pagination preceding that of the translation.

16. Maurice Merleau-Ponty, *L'oeil et l'esprit* (Paris: Gallimard, 1964). Translated by Michael B. Smith and revised by Leonard Lawlor as "Eye and Mind," in *The Merleau-Ponty Reader*, ed. Ted Toadvine and Leonard Lawlor (Evanston, IL: Northwestern University Press, 2007), 351–78. Hereafter cited as *OE*, with the original language pagination preceding that of the translation.

17. This expansion is, in fact, the primary foundation for Abram's views concerning nature, which has prompted criticism from those who do not readily accept that nonhuman responses, and more specifically nonliving "responses," can fulfill the requirements of a dialogical or discourse ethic. See, for example, Steven Vogel, "The Silence of Nature," *Environmental Values* 15, no. 2 (2006): 145–71.

18. "Aseity" here is meant in the sense of not having a self-contained essence, the opposite of ipseity, not in its theological sense.

19. Smith translates "*Le Sentir*" as "Sense Experience," which is accurate as to the content but loses the connection to *sentiment* (feeling) and *sens*, which itself has at least three distinct senses (meaning/direction/sense, the physical senses, as well as the first person conjugation of *sentir*). "Sensation" might render the title more aptly, though it might give the (mistaken) impression of advancing an empirical theory of perception. For this reason, I agree entirely with Landes's using "Sensing" as a translation.

20. I would like to thank my colleague Paul Carelli for making me aware of the discussions surrounding the differences between the later and middle dialogues regarding participation. Though I seek to offer a rather general and neutral interpretation with respect to those debates, relying predominantly on the accounts from the *Republic* and *Phaedo,* I have also benefited from reading Alexander Nehamas's "Participation and Predication in Plato's Later Thought," *Review of Metaphysics* 36, no. 2 (1982): 343–74, in explaining Plato's views.

21. As an example of a view that argues for a more concrete understanding of the relation, R. E. Allen claims that the relationship is a causal one, but he doesn't offer any textual evidence for this interpretation. See Allen, "Participation and Predication in Plato's Middle Dialogues," *Philosophical Review* 69, no. 2 (1960): 162.

22. Maurice Merleau-Ponty, *Le visible et l'invisible* (Paris: Gallimard, 1964), 234. Translated by Alphonso Lingis as *The Visible and the Invisible* (Evanston, IL: Northwestern University Press, 1968), 183. Hereafter cited as *VI*, with the original language pagination preceding that of the translation.

23. One of Heidegger's clearer expressions of this idea comes in "What Is Metaphysics?": "'Pure Being and pure Nothing are therefore the same.' This proposition of Hegel's . . . is correct. Being and the nothing do belong together, not because both—from the point of view of the Hegelian concept of thought—agree in their indeterminateness and immediacy, but rather because being itself is essentially finite and manifests itself in the transcendence of a Dasein that is held out into the nothing." "What Is Metaphysics?" trans. David F. Krell, in *Pathmarks*, ed. William McNeill (New York: Cambridge University Press, 1998), 94–95.

24. Maurice Merleau-Ponty, *Notes de cours: 1959–1961*, ed. Stéphanie Ménasé (Paris: Gallimard, 1996), 103. Hereafter cited as *NC*.

25. Specifically, Merleau-Ponty makes heavy use of "On the Question of Being," *The Principle of Reason*, and *What Is Called Thinking?* in addition to *Introduction to Metaphysics*, which is the principle text to which he refers in the lectures.

26. Merleau-Ponty, in juxtaposing Sartre's and Heidegger's respective views of being and nothingness, cites Heidegger's "On the Question of Being" in order to show that "being is that upon which nothingness opens" and that "without doubt, presence supposes an inverse of nihilation, but also that nihilation is 'veiled presence'" (*NC*, 103).

27. I mention this so-called malaise here because it is the source of Sheehan's pessimism concerning the possibility of a nondominating relation to beings. Merleau-Ponty, too, fails to note Heidegger's self-description of his project as preparatory, but intuits precisely the direction that Heidegger was preparing for an indirect expression of being. I make the claim that Merleau-Ponty's philosophical aims are Heideggerian in this sense, hoping to contextualize the subtle criticisms that Merleau-Ponty levels at Heidegger's philosophy within his overall acceptance of Heidegger's restoration of the phenomenological project on the basis of thinking the truth of being we discussed in the last chapter. After all, Merleau-Ponty does admit that Heidegger has a positive conclusion: that there is a "thinking" that is in correspondence with what is to be thought, *Besinnung*, or mindfulness. I propose, then, that Merleau-Ponty sees himself as offering a *Besinnung*, which he understands as the search for an "Urgemein Stiftung" (a primordially common institution; *VI*, 233/182).

28. As mentioned earlier, "divergence" is rendering the French term *écart*, which has also been translated as "gap," "spread," "hiatus," or "deviation." All these translations approximate Merleau-Ponty's use of the word, but, for the sake of consistency, the word will be rendered here as "divergence" alone. The operant sense of *écart* derives from the verb *écarter*, "to disjoin" or "to separate." To anticipate the coming argument, if being is said to be that which is *gathered together* (*rassemblé*), this can happen only if beings are *separate*; but disjoining, negation, requires conversely that there be joints that can be articulated.

29. Lawrence Hass defines *écart* nicely and succinctly as a "separation-difference" that accounts for the difference of the perceiver from the perceived while at the same time maintaining the immanence of the perceiver in the perceived. Like Heidegger's and Sartre's nothingness, the *écart* "is the very working of openness in perceptual experience which creates the space or gap through which sensibility occurs." On this understanding, *écart* is the means by which beings are differentiated at their pivots or

joints. See Hass, *Merleau-Ponty's Philosophy* (Bloomington: Indiana University Press, 2008), 129–32.

30. *Côté* is a technical term for Merleau-Ponty that he, in turn, receives from Proust, as found in the titles of two of his books, *Du côté de chez Swann* and *Le côté de Guermantes*, usually translated as *Swann's Way* and *The Guermantes Way*, respectively. The use of the word is dual, and both senses are included in Merleau-Ponty's own use. On the one hand, *côté* is intended as a perspectival word: how things appear for or according to Swann. On the other hand, *côté* is intended in the literal sense of "path," as in Heidegger's *Weg*: from Proust's narrator's childhood home, his family would take walks along either a path that passed by Swann's home or that of the Guermanteses. In saying that the mind is a *côté* of the body, then, Merleau-Ponty is not saying that the mind actually has a physical place inside the body, but that from our perspective as the sensing body, we have a mind.

31. I compare the sense of flesh I am about to describe to the currently accepted interpretations of the concept in greater detail in the second and third sections of my essay "Flesh and Nature."

32. On this point, see also *VI*, 138/103, where Merleau-Ponty describes both natural and historical institutions as dimensions that act as standards of measure. I translate *mesurant* with "standard of measure" rather than with Lingis's "measurants" because of the connection that Merleau-Ponty makes between the *mesurant* and Husserl's "norming [*Normierung*]" (*VI*, 246/196).

33. Even if one correctly points out that animals, on Heidegger's account, are poor in world rather than lacking in world like the stone, it remains the case that their being "captivated" in their relations to other beings prevents them from dealing with objects as such, from projecting open being. I explain the difference between Merleau-Ponty and Heidegger on this point in "Animals, Language and Life: Searching for Animal Attunement with Heidegger and Merleau-Ponty," *Environmental Philosophy* 6, no. 1 (2009): 21–34, but Buchanan's study of their two positions with respect to Uexküll's views, from which they are derived in *Onto-Ethologies*, is much more thorough and to the point.

34. Maurice Merleau-Ponty, *The Merleau-Ponty Reader*, ed. Ted Toadvine and Leonard Lawlor (Evanston, IL: Northwestern University Press, 2007), 421.

35. Alfred North Whitehead, *The Concept of Nature: Tarner Lectures Delivered in Trinity College, November 1919* (Cambridge: Cambridge University Press, 1920). Hereafter cited as *CN*. In making the above claim, I refer to the following passage:

> It is an exhibition of the process of nature that each duration happens and passes. The process of nature can also be termed the passage of nature. I definitively refrain at this stage from using the word "time," since the measurable time of civilized life generally merely exhibits some aspects of the more fundamental fact of the passage of nature. I believe that in this doctrine I am in full accord with Bergson, though he uses "time" for the fundamental fact which I call the "passage of nature." Also the passage of nature is exhibited in spatial transition as well as temporal transition. (*CN*, 54)

36. I also restrict myself to commenting on texts prior to the introduction of "god" into Whitehead's system as the principle of limitation in Whitehead, *Science and the Modern World* (New York: Free Press, 1925), esp. chap. 10–12.

37. Merleau-Ponty is apparently utilizing the French translation of "simple location" as "*emplacement unique*," which Vallier translates back into English as "unique emplacement" (*N*, 154n2/113).

38. Whitehead distinguishes between events and physical objects, which are defined as recognizable characters of events: "You cannot recognize an event; because when it is gone, it is gone. You may observe another event of analogous character, but the actual chunk of the life of nature is inseparable from its unique occurrence. But a character of an event can be recognized. . . . Things which we thus recognize I call objects" (*CN*, 169). Physical objects, then, are not constantly present beings that exist independently of the events within which they manifest as in the metaphysical tradition. Rather, physical objects are qualitative similarities recognizable between distinct events of manifestation, making them intellectual abstractions from a series of events rather than entities themselves (189–90).

39. For more on the distinction between human signification and natural signification, see the "second sketch" of Merleau-Ponty's third course on nature (*N*, 278–83/216–20). Natural communication is described as the flesh of the world, with expressive gestures developing within that flesh another "hollow" or dimension that is of that flesh but at the same time is invisible within it (cf. esp. 281–82/219).

40. This is the limitation of Toadvine's book on Merleau-Ponty's conception of nature, which does not reckon with Whitehead's influence on Merleau-Ponty's views in even a cursory way. Though I concur with most of Toadvine's accounts of Merleau-Ponty's basic concepts such as expression and the conception of phenomenology as "recognizing reflection's ineliminable debt to a prereflective history that always exceeds its reach" (*Merleau-Ponty's Philosophy of Nature*, 122), I also find that by adhering to the traditional view that the foundation of Merleau-Ponty's philosophy is perceptual impedes his inquiry from exploring some of the more interesting consequences of the theory of flesh that I will discuss below. Situating Merleau-Ponty's philosophy as a speculative process philosophy, for example, allows us to move beyond the conception of nature's expression simply as the "good error" of formalizing silent nature in human meaningful expressions (126–30). As we will see, there is a sense in which the process interpretation of flesh as the being of beings reveals a nature as genuinely expressive and generative. The consequence of Toadvine's remaining ensconced within the more traditional project of phenomenology is that the silence of nature that is not the contrary to language in Merleau-Ponty's thought is limited in its importance to how it can be *brought* to expression in an instituted human signifying system. I do not intend to imply here that Toadvine is inaccurate in his interpretation of Merleau-Ponty; there is ample evidence, which he supplies, that Merleau-Ponty himself still agreed with certain aspects of the traditional view. My suggestion, then, can be understood as either that we need to move beyond a strict interpretation of Merleau-Ponty's phenomenology in order to realize the potential of the relational ontology Merleau-Ponty suggests in his *unfinished* final philosophy (if my reader finds Toadvine's interpretation accurate) *or* that we need to reinterpret some of the comments Merleau-Ponty makes in light of his adoption of process philosophy. In either case, Toadvine's book leaves us with any number of questions, ranging from how the mode of natural expression he describes at the end of the text differs from the Kantian assertion that no *certain knowledge* is possible of the world as it really is, to what nature is actually like in Merleau-Ponty's philosophy as opposed to simply what position it holds and what role it fulfills within the ontology in general.

41. One of the ways in which the account Toadvine gives can be developed further is by incorporating this insight into multiple relation. All his examples of chiasm are between two terms, which leads him when discussing the chiasmus to miss the mediating role of the fifth term of the relation described above.

42. Isabelle Stengers has done an admirable job of this exact task in her *Thinking with Whitehead: A Free and Wild Creation of Concepts*, trans. Michael Chase (Cambridge, MA: Harvard University Press, 2011). Stengers's work does an excellent job of showing the continuity between Whitehead's process philosophy and more contemporary speculative philosophies such as Latour's and Deleuze's. I note this connection because, though she claims Whitehead is no phenomenologist (105), which is true in the sense she describes in her text, it remains the case that the way she describes the value of Whitehead's philosophy as animating an original experience of the world and respecting that experience throughout one's speculative engagement with it is, on the view developed in this work, a profoundly phenomenological project. Consider, for example, the following: "The Whiteheadean scheme is a work of politeness because it seeks to be 'adequate,' refraining from insulting any living value. Yet if it is to be effective, it owes this to its fabrication, to the articulation of constraints that oblige thought, an 'asignificant' matrix whose truth is the events of 'disclosure' it may induce, modifying a being's relations to its environment, and, in this case, the relations of a statement to its consequences" (518).

43. In many ways, Merleau-Ponty's conception of the ensemble is a precursor to the idea of an assemblage as it is found in both Deleuze's work and the science of ecology.

44. Whitehead does advance the notion of an "event-particle," which he defines as an "ideal minimum limit to events," but this limit is itself an abstraction from an event and thereby "is a group of abstractive sets" and not a thing in the traditional sense (*CN*, 86). Even if one were to argue, then, that there is a smallest possible event, this event is still not homogeneous as in the conception of matter, but is a heterogeneous set of elements. Additionally, what Whitehead has in mind when he speaks of an "abstractive set of elements" is not microscopic particles, but rather "a mere spatial point-flash of instantaneous duration" (172). An analogy might be viewing a film frame by frame. In this case, each frame would capture an event-particle, but given that there is a temporal elapse between the two frames, no matter how close together they become, no frame taken independently can express the total movement of passage of the film because there is nothing relative to which direction can be indicated.

45. Stengers, *Thinking with Whitehead*, 508.

46. Stengers makes a strong case for identifying Whitehead's conception of "creativity" with the plane of immanence in Deleuze's thought based upon Whitehead's claim in *Process and Reality* that creativity is intended to replace the traditional category of matter. It is on this basis that we can draw Latour even closer to Merleau-Ponty insofar as nature, as the flesh of the world, can then be said to possess its own generativity, its own productive potential, rather than an inherent and preestablished order.

47. John Sallis, *Force of Imagination: The Sense of the Elemental* (Bloomington: Indiana University Press, 2000). Hereafter cited as *FI*.

48. The term "experimental metaphysics" is Latour's from *The Politics of Nature*.

49. This point is a strength of Toadvine's interpretation. In discussing the open-ended nature of divergence, he is able to explicate the human-nature relationship as one

of mutual becoming, wherein the understanding of each term affects the understanding of the other rather than an oppositional definition. *Merleau-Ponty's Philosophy of Nature*, 109, 116.

50. Maurice Merleau-Ponty, *Les aventures de la dialectique*, folio ed. (Paris: Gallimard, 1955), 282. Translated by Joseph Bien as *Adventures of the Dialectic* (Evanston, IL: Northwestern University Press, 1973), 204. Hereafter cited as *AD*, with the original language pagination preceding that of the translation.

51. The same is implied in the discussion of the "non-communist left's" role in society. Merleau-Ponty chastises communism for "renouncing a harassing action" (*AD*, 314/227) and claims that social progress will not be made through revolutionary movements, which will be legitimated as adversaries to the established institutions as "a useful menace, as a continual call to order," but rather through "a conscious action which will confront itself with the judgment of an opposition" (312/226).

52. One cannot help but also hear a reference to Whitehead's *Adventures of Ideas*. Stengers also describes Whitehead's philosophical project itself as an adventure in the sense described here.

53. I will use a translation of *Wahlverwandschaft* closer to the French rendering, "chosen kinship," in order to best show the continuity of Merleau-Ponty's thinking.

54. It may seem as if the interpretation offered deviates from Merleau-Ponty's aforementioned claim that nature is the noninstituted. On the one hand, it has: nature is a collectively constructed common world, continually being reinstituted through the changes in its internal relations. On the other hand, in the sense Barbaras described, nature remains the noninstituted: it is the wild being encountered in brute perception. So, if one is speaking from a more traditional phenomenological perspective, nature is noninstituted. If, however, one shifts phenomenology in the direction of a general theory of affectivity as I have attempted to do here, nature becomes an instituted phenomenon.

55. For example, he states the following: "'Humanism' now means, in case we decide to retain the word, that the essence of the human being is essential for the truth of being, specifically in such a way that what matters is not the human being simply as such. So we are thinking a curious kind of 'humanism.'" Martin Heidegger, "Brief über den 'Humanismus,'" in *Gesamtausgabe*, vol. 4, *Wegmarken* (Frankfurt: Klostermann, 1967), 342. Translated by Frank A. Capuzzi as "Letter on Humanism," in *Pathmarks*, ed. William McNeill (New York: Cambridge University Press, 1998), 263.

56. Heidegger, "Letter on Humanism," 156/247.

CONCLUSION

1. Val Plumwood, *Environmental Culture: The Ecological Crisis of Reason* (New York: Routledge, 2002), 124.

2. Ibid., 128–29.

3. John Stuart Mill, *Utilitarianism* (Indianapolis, IN: Hackett, 2001), 32–34.

4. Rather than discuss this possibility speculatively here, I will look forward to the forthcoming publication of *Enquête sur les modes d'existence* and the prospective publication of his Gifford lectures on "Facing Gaia," both of which I anticipate will shed greater light on this issue.

5. I mentioned William McNeill's excellent interpretation of Heidegger's views on this matter earlier ("Life Beyond the Organism: Animal Being in Heidegger's Freiburg

Lectures, 1929–30," in *Animal Others: On Ethics, Ontology, and Animal Life*, ed. H. Peter Steeves [Albany: State University of New York Press, 1999], 197–248). Though McNeill argues that Heidegger's views are not humanistic in the traditional sense of offering a view of animals that allows them to be thought of as "present at hand" or there for the utilization of human beings (232), he still acknowledges the divergence between the "captivation" of the animal and the "world-building" of the human being, which allows human beings to "let beings be" in a way that the animal cannot in its captivation. The affirmation of human ethicality that this sanctions (242) is made on the basis of the presumed human privileged relation to being. Thus, I am in effect agreeing with McNeill's characterization of Heidegger's position while at the same time proposing that the charges Derrida levels at Heidegger in *Of Spirit* and the last chapter of *L'animal que donc je suis* concerning his latent humanism still have validity.

6. Toadvine cites the one place where Merleau-Ponty questions this language of strata as a definitive rejection of "the juxtaposing metaphors of 'leave' and 'layers' for describing the sensible and sentient dimensions of the body" (*Merleau-Ponty's Philosophy of Nature*, 115). I don't see a contradiction between our two interpretations, however, given that he is speaking of a relation *to the body itself*, whereas I am discussing strata as occurring within the world *between beings*. So where the auto-affective touch can be "one sole circular course," this course is multiplied infinitely when taking account of the multiple relationality discussed in chapter 4.

7. This locution in which the "obverse-reverse" structure of the leaf of paper is combined with the language of layers and strata is more evidence that the contrast is not as stark as Toadvine would suggest.

8. In working out an account of place, I have specifically in mind the work of Edward S. Casey, who has been writing on the subject for the past fifteen years. His books of note in this regard are *Getting Back into Place: Toward a Renewed Understanding of the Place-World* (Bloomington: Indiana University Press, 1993); *Representing Place: Landscape Painting and Maps* (Minneapolis: University of Minnesota Press, 2002); *The Fate of Place: A Philosophical History* (Berkeley: University of California Press, 1998); and, more recently, *The World at a Glance* (Bloomington: Indiana University Press, 2007). Since I cannot hope here in the conclusion of this work to match the erudition of his account that takes place over the course of these lengthy texts, I will at this point simply indicate that his views on place are far more developed than my own and refer my reader to his work, especially *Getting Back into Place*, chap. 7 and 8; and *The World at a Glance*, part 4, especially chap. 11. Plumwood has also extensively commented on the need for an ethics of place in *Environmental Culture*, and such an emphasis is the explicit premise behind movements such as bioregionalism (e.g., Robert L. Thayer, Jr., *LifePlace: Bioregional Thought and Practice* [Berkeley: University of California Press, 2003]).

9. Georges Canguilhem, *The Normal and the Pathological*, trans. Carolyn R. Fawcett and Robert S. Cohen (New York: Zone Books, 1989). Hereafter cited as *NP*.

BIBLIOGRAPHY

Abram, David. "Merleau-Ponty and the Voice of the Earth." *Environmental Ethics* 10 (1988): 101–20.

———. *The Spell of the Sensuous: Perception and Language in a More-Than-Human World*. New York: Vintage, 1996.

Allen, R. E. "Participation and Predication in Plato's Middle Dialogues." *Philosophical Review* 69, no. 2 (1960): 147–64.

Aristotle. *The Metaphysics*. Bks. 1–9. Translated by Hugh Tredennick. Cambridge, MA: Harvard University Press, 1956.

———. *The Physics*. Vol. 1, bks. 1–4. Translated by Philip H. Wicksteed and Francis M. Cornford. Cambridge, MA: Harvard University Press, 1957.

Bannon, Bryan E. "Animals, Language, and Life: Searching for Animal Attunement with Heidegger and Merleau-Ponty." *Environmental Philosophy* 6, no. 1 (2009): 21–34.

———. "Developing Val Plumwood's Dialogical Ethical Ontology and Its Consequences for a Place-Based Ethic." *Ethics and the Environment* 14, no. 2 (2009): 39–55.

———. "Flesh and Nature: Understanding Merleau-Ponty's Relational Ontology." *Research in Phenomenology* 41 (2011): 327–57.

Barbaras, Renaud. "Merleau-Ponty et la nature." *Chiasmi International* 2 (2000): 47–61.

Bennett, Jane. *Vibrant Matter: A Political Ecology of Things*. Durham, NC: Duke University Press, 2010.

Bookchin, Murray. "Social Ecology Versus 'Deep' Ecology: A Challenge to the Ecology Movement." *Green Perspectives: Newsletter of the Green Program Project* 4–5 (1987): 1–23.

———. "Toward a Philosophy of Nature—The Bases for an Ecological Ethics." In *Deep Ecology*, edited by Michael Tobias, 213–39. San Diego: Avant Books: 1985.

Brown, Charles S. "The Real and the Good: Phenomenology and the Possibility of an Axiological Rationality." In Brown and Toadvine, *Eco-Phenomenology*, 3–18.

Brown, Charles S., and Ted Toadvine, eds. *Eco-Phenomenology: Back to the Earth Itself*. Albany: State University of New York Press, 2003.

Buchanan, Brett. *Onto-Ethologies: The Animal Environments of Uexküll, Heidegger, Merleau-Ponty, and Deleuze*. Albany: State University of New York Press, 2008.

Canguilhem, Georges. *The Normal and the Pathological*. Translated by Carolyn R. Fawcett and Robert S. Cohen. New York: Zone Books, 1989.

Casey, Edward S. *The Fate of Place: A Philosophical History*. Berkeley: University of California Press, 1998.

———. *Getting Back into Place: Toward a Renewed Understanding of the Place-World*. Bloomington: Indiana University Press, 1993.

———. *Representing Place: Landscape Painting and Maps*. Minneapolis: University of Minnesota Press, 2002.

———. "Taking a Glance at the Environment." In Brown and Toadvine, *Eco-Phenomenology*, 187–210.

———. *The World at a Glance*. Bloomington: Indiana University Press, 2007.

Cooper, David E. "Heidegger on Nature." *Environmental Values* 14, no. 3 (2005): 339–51.

Crist, Eileen. "Against the Social Construction of Nature and Wilderness." *Environmental Ethics* 26 (2004): 5–24.

Curry, Patrick. "Re-Thinking Nature: Towards an Eco-Pluralism." *Environmental Values* 12, no. 3 (2003): 337–60.

Deleuze, Gilles, and Félix Guattari. *A Thousand Plateaus: Capitalism and Schizophrenia*. Translated by Brian Massumi. Minneapolis: University of Minnesota Press, 1987.

Derrida, Jacques. *L'animal que donc je suis*. Paris: Galilée, 2006.

———. *Of Spirit*. Translated by Geoffrey Bennington and Rachel Bowlby. Chicago: University of Chicago Press, 1989.

Drenthen, Martin. "Reading Ourselves through the Land: Landscape Hermeneutics and Ethics of Place." In *Placing Nature on the Borders of Religion, Philosophy and Ethics*, edited by Forrest Clingerman and Mark H. Dixon, 123–38. Farnham: Ashgate, 2011.

Evernden, Neil. *The Social Creation of Nature*. Baltimore: Johns Hopkins University Press, 1992.

Foltz, Bruce V. *Inhabiting the Earth: Heidegger, Environmental Ethics, and the Metaphysics of Nature*. Atlantic Highlands, NJ: Humanities Press, 1995.

Foucault, Michel. "The Ethics of the Concern for the Self as a Practice of Freedom." In *Ethics: Subjectivity and Truth*, edited by Paul Rabinow, 281–301. New York: New Press, 2006.

Fox, Warwick. *Toward a Transpersonal Ecology: Developing New Foundations for Environmentalism*. Albany: State University of New York Press, 1995.

Glazebrook, Trish. *Heidegger's Philosophy of Science*. New York: Fordham University Press, 2000.

Hadot, Pierre. *The Veil of Isis: An Essay on the History of the Idea of Nature*. Translated by Michael Chase. Cambridge, MA: Belknap, 2006.

Harman, Graham. "The Importance of Bruno Latour for Philosophy." *Cultural Studies Review* 13, no. 1 (2007): 31–49.

———. *Prince of Networks: Bruno Latour and Metaphysics*. Melbourne: re.press, 2009.

Hass, Lawrence. *Merleau-Ponty's Philosophy*. Bloomington: Indiana University Press, 2008.

Hayward, Tim. "Anthropocentrism: A Misunderstood Problem." *Environmental Values* 6, no. 1 (1997): 49–63.

Heidegger, Martin. "Brief über den'Humanismus.'" In *Gesamtausgabe*, vol. 4, *Wegmarken*, 311–60. Translated by Frank A. Capuzzi as "Letter on 'Humanism,'" in *Pathmarks*, 239–76.

———. "Der Ursprung des Kunstwerkes." In *Holzwege*, 7–68. Translated by Julian Young and Kenneth Haynes as "The Origin of the Work of Art," in *Off the Beaten Track*, 1–56.

———. "Die Frage nach der Technik." In *Gesamtausgabe*, vol. 7, *Vorträge und Aufsätze*, 7–36. Frankfurt: Klostermann, 2000. Translated by William Lovitt as "The Question Concerning Technology," in *The Question Concerning Technology and Other Essays* (San Francisco: Harper Colophon, 1977), 3–35.

———. "Die Zeit des Weltbildes." In *Holzwege*, 69–104. Translated by Julian Young and Kenneth Haynes as "The Age of the World Picture," in *Off the Beaten Track*, 57–85.

———. *Gesamtausgabe*. Vol. 4, *Wegmarken*. Frankfurt: Klostermann, 1967.

———. "Memorial Address." In *Discourse on Thinking*, 43–57. New York: Harper & Row, 1966.

———. *Gesamtausgabe*. Vol. 6, *Nietzsche I*. Frankfurt: Klostermann, 1996. Translated by David F. Krell as *Nietzsche*, vol. 1–2 (San Francisco: HarperCollins, 1991).

———. *Gesamtausgabe*. Vol. 29–30, *Die Grundbegriffe der Metaphysik: Welt—Endlichkeit—Einsamkeit*. Frankfurt: Klostermann, 1983. Translated by William McNeill and Nicholas Walker as *The Fundamental Concepts of Metaphysics: World, Finitude, Solitude* (Indianapolis: Indiana University Press, 1995).

———. *Gesamtausgabe*. Vol. 40, *Einführung in die Metaphysik*. Frankfurt: Klostermann, 1983. Translated by Gregory Fried and Richard Polt as *Introduction to Metaphysics* (New Haven, CT: Yale University Press, 2000).

———. *Gesamtausgabe*. Vol. 65, *Beiträge zur Philosophie*. Frankfurt: Klostermann, 1989. Translated by Parvis Emad and Kenneth Maly as *Contributions to Philosophy (from Enowning)* (Bloomington: Indiana University Press, 1999).

———. *Gesamtausgabe*. Vol. 66, *Besinnung*. Frankfurt: Klostermann, 1997. Translated by Parvis Emad and Thomas Kalary as *Mindfulness* (London: Continuum, 2006), 257–58.

———. "Hegels Begriff der Erfahrung." In *Holzwege*, 115–208. Translated by Julian Young and Kenneth Haynes as "Hegel's Concept of Experience," in *Off the Beaten Track*, 86–156.

———. *Holzwege*. Frankfurt: Klostermann, 1957.

———. *Off the Beaten Track*. Edited and translated by Julian Young and Kenneth Haynes. New York: Cambridge University Press, 2002.

———. *Pathmarks*. Edited by William McNeill. New York: Cambridge University Press, 1998.

———. *Sein und Zeit*. Tübingen: Niemeyer, 1927. Translated by Joan Stambaugh as *Being and Time* (Albany: State University of New York Press, 1996).

———. "Vom Wesen und Begriff der Physis Aristoteles, *Physik B, 1*." In *Gesamtausgabe*, vol. 4, *Wegmarken*, 237–99. Translated by Thomas Sheehan as "On the Essence and Concept of *Physis* in Aristotle's *Physics* B, I," in *Pathmarks*, 183–230.

———. "What Is Metaphysics?" Translated by David F. Krell. In *Pathmarks*, 82–96.

Heyd, Thomas, ed. *Recognizing the Autonomy of Nature: Theory and Practice*. New York: Columbia University Press, 2005.

Husserl, Edmund. *Ideas Pertaining to a Pure Phenomenology and to a Phenomenological Philosophy*. First Book: *General Introduction to a Pure Phenomenology*. Translated by Fred Kersten. The Hague: Martinus Nijhoff, 1983.

———. *Ideas Pertaining to a Pure Phenomenology and to a Phenomenological Philosophy.* Second Book: *Studies in the Phenomenology of Constitution.* Translated by Richard Rojcewicz and André Schuwer. Dordrecht: Kluwer, 1989.

Ihde, Don. *Heidegger's Technologies: Postphenomenological Perspectives.* New York: Fordham University Press, 2010.

Khong, Lynnette. "Actants and Enframing: Heidegger and Latour on Technology." *Studies in the History and Philosophy of Science* 34 (2003): 693–704.

Kidner, David W. "Fabricating Nature: A Critique of the Social Construction of Nature." *Environmental Ethics* 22 (2000): 339–57.

Klaver, Irene J. "Boundaries on the Edge." In *Nature's Edge: Boundary Explorations in Ecological Theory and Practice*, edited by Charles S. Brown and Ted Toadvine, 146–76. Albany: State University of New York Press, 2006.

Kochan, Jeff. "Latour's Heidegger." *Social Studies of Science* 40, no. 4 (2010): 579–98.

Latour, Bruno. *Pandora's Hope: Essays on the Reality of Science Studies.* Cambridge, MA: Harvard University Press, 1999.

———. *The Pasteurization of France.* Translated by Alan Sheridan and John Law. Cambridge, MA: Harvard University Press, 1988.

———. *Politics of Nature: How to Bring the Sciences into Democracy.* Translated by Catherine Porter. Cambridge, MA: Harvard University Press, 2004.

———. *We Have Never Been Modern.* Translated by Catherine Porter. Cambridge, MA: Harvard University Press, 1993.

Leibniz, G. W. F. "Letter to Arnauld (April 30, 1687)." Translated by Roger Ariew and Daniel Garber. In *Philosophical Essays*, 81–90. Indianapolis, IN: Hackett, 1989.

Leopold, Aldo. *A Sand County Almanac.* New York: Ballantine, 1970.

Lovelock, James. *Gaia: A New Look at Life on Earth.* New York: Oxford University Press, 2000.

———. *The Revenge of Gaia: Earth's Climate Crisis and the Fate of Humanity.* New York: Basic Books, 2007.

Marratto, Scott L. *The Intercorporeal Self: Merleau-Ponty on Subjectivity.* Albany: State University of New York Press, 2012.

McNeill, William. "Life Beyond the Organism: Animal Being in Heidegger's Freiburg Lectures, 1929–30." In *Animal Others: On Ethics, Ontology, and Animal Life*, edited by H. Peter Steeves, 197–248. Albany: State University of New York Press, 1999.

Meillassoux, Quentin. *After Finitude: An Essay on the Necessity of Contingency.* Translated by Ray Brassier. New York: Continuum, 2010.

Merchant, Carolyn. *Reinventing Eden: The Fate of Nature in Western Culture.* New York: Routledge, 2004.

Merleau-Ponty, Maurice. *La nature: Notes, cours du Collège de France.* Paris: Seuil, 1994. Translated by Robert Vallier as *Nature: Course Notes from the Collège de France* (Evanston, IL: Northwestern University Press, 2003).

———. *La prose du monde.* Edited by Claude Lefort. Paris: Gallimard, 1969. Translated by John O'Neill as *The Prose of the World* (Evanston, IL: Northwestern University Press, 1973).

———. *La structure du comportement.* Paris: Quadridge/PUF, 2006. Translated by Alden Fisher as *The Structure of Behavior* (Pittsburgh, PA: Duquesne University Press, 1963).

———. "Le langage indirect et les voix du silence." In *Signes*, folio ed., 63–135. Paris: Gallimard, 1960. Translated by Michael B. Smith and revised by Leonard Lawlor as "Indirect Language and the Voices of Silence," in *The Merleau-Ponty Reader*, 241–82.

———. *Les aventures de la dialectique*. Folio ed. Paris: Gallimard, 1955. Translated by Joseph Bien as *Adventures of the Dialectic* (Evanston, IL: Northwestern University Press, 1973).

———. *Le visible et l'invisible*. Paris: Gallimard, 2003. First published 1964. Translated by Alphonso Lingis as *The Visible and the Invisible* (Evanston, IL: Northwestern University Press, 1968).

———. *L'oeil et l'ésprit*. Paris: Gallimard, 1964. Translated by Michael B. Smith and revised by Leonard Lawlor as "Eye and Mind," in *The Merleau-Ponty Reader*, 351–78.

———. *The Merleau-Ponty Reader*. Edited by Ted Toadvine and Leonard Lawlor. Evanston, IL: Northwestern University Press, 2007.

———. *Notes de cours: 1959–1961*. Edited by Stéphanie Ménasé. Paris: Gallimard, 1996.

———. *Phénoménologie de la perception*. Paris: Gallimard, 1945. Translated by Colin Smith as *Phenomenology of Perception* (New York: Routledge, 1962).

Mill, John Stuart. *Utilitarianism*. Indianapolis, IN: Hackett, 2001.

Naess, Arne. *Ecology, Community and Lifestyle: Outline of an Ecosophy*. Translated and edited by David Rothenberg. Cambridge: Cambridge University Press, 1993.

Nash, Roderick. *Wilderness and the American Mind*. 4th ed. New Haven, CT: Yale University Press, 2001.

Nehamas, Alexander. "Participation and Predication in Plato's Later Thought." *Review of Metaphysics* 36, no. 2 (1982): 343–74.

Noë, Alva. *Action in Perception*. Cambridge, MA: MIT Press, 2004.

Norton, Bryan G. "Environmental Ethics and Weak Anthropocentrism." *Environmental Ethics* 6 (1984): 131–48.

———. *Toward Unity Among Environmentalists*. New York: Oxford University Press, 1991.

Oelschlaeger, Max. *The Idea of Wilderness: From Prehistory to the Age of Ecology*. New Haven, CT: Yale University Press, 1993.

O'Regan, J. Kevin, and Alva Noë. "A Sensorimotor Account of Vision and Visual Consciousness." *Behavioral and Brain Sciences* 24, no. 5 (2001): 939–73.

Plumwood, Val. *Environmental Culture: The Ecological Crisis of Reason*. New York: Routledge, 2002.

———. *Feminism and the Mastery of Nature*. New York: Routledge, 1993.

———. "Toward a Progressive Naturalism." In Heyd, *Recognizing the Autonomy of Nature*, 25–53.

Regan, Tom. "The Nature and Possibility of an Environmental Ethic." *Environmental Ethics* 3 (1981): 19–34.

Riis, Søren. "The Symmetry Between Bruno Latour and Martin Heidegger: The Technique of Turning a Police Officer into a Speed Bump." *Social Studies of Science* 38, no. 2 (2008): 285–301.

Routley, Richard. "Is There a Need for a New, an Environmental, Ethic?" *Proceedings, 15th World Congress of Philosophy* 1 (1973): 205–10.

Sallis, John. *Force of Imagination: The Sense of the Elemental*. Bloomington: Indiana University Press, 2000.

Schelling, F. W. J. *First Outline of a System of the Philosophy of Nature*. Translated by Keith R. Peterson. Albany: State University of New York Press, 2004.

Schiølin, Kasper. "Follow the Verbs! A Contribution to the Study of the Heidegger-Latour Connection." *Social Studies of Science* 42, no. 5 (2012): 775–86.

Sessions, George. "Postmodernism and Environmental Justice: The Demise of the Ecology Movement?" *Trumpeter* 12, no. 3 (1995): 150–54.

———. "Reinventing Nature . . . ? A Response to William Cronon's *Uncommon Ground*." *Trumpeter* 13, no. 1 (1996): 33–38.

Sheehan, Thomas. "Nihilism: Heidegger/Jünger/Aristotle." In *Phenomenology: Japanese and American Perspectives*, edited by Burt C. Hopkins, 273–316. Dordrecht: Kluwer, 1998.

Soulé, Michael E. "The Social Siege of Nature." In *Reinventing Nature? Responses to Postmodern Deconstruction*, edited by Michael E. Soulé and Gary Lease, 137–70. Washington, DC: Island Press, 1995.

Stengers, Isabelle. *Thinking with Whitehead: A Free and Wild Creation of Concepts*. Translated by Michael Chase. Cambridge, MA: Harvard University Press, 2011.

Thayer, Robert L., Jr. *LifePlace: Bioregional Thought and Practice*. Berkeley: University of California Press, 2003.

Toadvine, Ted. *Merleau-Ponty's Philosophy of Nature*. Evanston, IL: Northwestern University Press, 2009.

Van Buren, John. "Critical Environmental Hermeneutics." *Environmental Ethics* 17 (1995): 259–75.

Varela, Francisco J. "The Specious Present: A Neurophenomenology of Time Consciousness." In *Naturalizing Phenomenology: Issues in Contemporary Phenomenology and Cognitive Science*, edited by Jean Petitot, Francisco J. Varela, Bernard Pachoud, and Jean-Michel Roy, 266–306. Stanford, CA: Stanford University Press, 1999.

Vogel, Steven. *Against Nature: The Concept of Nature in Critical Theory*. Albany: State University of New York Press, 1996.

———. "Environmental Philosophy after the End of Nature." *Environmental Ethics* 24 (2002): 23–39.

———. "The Silence of Nature." *Environmental Values* 15, no. 2 (2006): 145–71.

White, Lynn, Jr. "The Historical Roots of Our Ecologic Crisis." *Science* 155 (1967): 1203–7.

Whitehead, Alfred North. *The Concept of Nature: Tarner Lectures Delivered in Trinity College, November 1919*. Cambridge: Cambridge University Press, 1920.

———. *Science and the Modern World*. New York: Free Press, 1925.

Williams, Bernard. *Shame and Necessity*. Berkeley: University of California Press, 1993.

Wood, David. "What Is Eco-Phenomenology?" In Brown and Toadvine, *Eco-Phenomenology*, 211–33.

Zimmerman, Michael E. "Beyond Humanism: Heidegger's Understanding of Technology." *Listening* 12 (1977): 74–83.

———. *Contesting Earth's Future: Radical Ecology and Postmodernity*. Berkeley: University of California Press, 1997.

———. "Heidegger's Phenomenology and Contemporary Environmentalism." In Brown and Toadvine, *Eco-Phenomenology*, 73–101.

———. "Implications of Heidegger's Thought for Deep Ecology." *Modern Schoolman* 64 (1986): 19–43.

———. "Toward a Heideggerian *Ethos* for Radical Environmentalism." *Environmental Ethics* 5 (1983): 99–131.

INDEX